Darwin, and After Darwin (annotated)

by George John Romanes

PREFACE

Several years ago Lord Rosebery founded, in the University of Edinburgh, a lectureship on "The Philosophy of Natural History," and I was invited by the Senatus to deliver the lectures. This invitation I accepted, and subsequently constituted the material of my lectures the foundation of another course, which was given in the Royal Institution, under the title "Before and after Darwin." Here the course extended over three years--namely from 1888 to 1890. The lectures for 1888 were devoted to the history of biology from the earliest recorded times till the publication of the "Origin of Species" in 1859; the lectures for 1889 dealt with the theory of organic evolution up to the date of Mr. Darwin's death, in 1882; while those of the third year discussed the further developments of this theory from that date till the close of the course in 1890.

It is from these two courses--which resembled each other in comprising between thirty and forty lectures, but differed largely in other respects--that the present treatise has grown. Seeing, however, that it has grown much beyond the bulk of the original lectures, I have thought it desirable to publish the whole in the form of three separate works. Of these the first--or that which deals with the purely historical side of biological science--may be allowed to stand over for an indefinite time. The second is the one which is now brought out and which, as its sub-title signifies, is devoted to the general theory of organic evolution as this was left by the stupendous labours of Darwin. As soon as the translations shall have been completed, the third portion will follow (probably in the Autumn season), under the sub-title, "Post-Darwinian Questions."

As the present volume is thus intended to be merely a systematic exposition of what may be termed the Darwinism of Darwin, and as on this account it is likely to prove of more service to general readers than to professed naturalists, I have been everywhere careful to avoid assuming even the most elementary knowledge of natural science on the part of those to whom the exposition is addressed. The case, however, will be different as regards the next volume, where I shall have to deal with the important questions touching Heredity, Utility, Isolation, &c., which have been raised since the death of Mr. Darwin, and which are now being debated with such salutary vehemence by the best naturalists of our time.

My obligations to the Senatus of the University of Edinburgh, and to the Board of Management of the Royal Institution, have already been virtually expressed; but I should like to take this opportunity of also expressing my obligations to the students who attended the lectures in the University of Edinburgh. For alike in respect of their large numbers, their keen intelligence, and their generous sympathy, the members of that voluntary class yielded a degree of stimulating encouragement, without which the labour of preparing the original lectures could not have been attended with the interest and the satisfaction that I found in it. My thanks are also due to Mr. R. E. Holding for the painstaking manner in which he has assisted me in executing most of the original drawings with which this volume is illustrated; and likewise to Messrs. Macmillan and Co. for kindly allowing me to reprint--without special acknowledgment in every case--certain passages from an essay which they published for me many years ago, under the title "Scientific Evidences of Organic Evolution." Lastly, I must mention that I am indebted to the same firm for permission to reproduce an excellent portrait of Mr. Darwin, which constitutes the frontispiece.

G. J. R.

CHRIST CHURCH, OXFORD, April 19th, 1892.

CONTENTS

SECTION I

EVOLUTION

CHAPTER I.

INTRODUCTORY.

Among the many and unprecedented changes that have been wrought by Mr. Darwin's work on the Origin of Species, there is one which, although second in importance to no other, has not received the attention which it deserves. I allude to the profound modification which that work has produced on the ideas of naturalists with regard to method.

Having had occasion of late years somewhat closely to follow the history of biological science, I have everywhere observed that progress is not so much marked by the march of discovery per se, as by the altered views of method which the march has involved. If we except what Aristotle called "the first start" in himself, I think one may fairly say that from the rejuvenescence of biology in the sixteenth century to the stage of growth which it has now reached in the nineteenth, there is a direct proportion to be found between the value of work done and the degree in which the worker has thereby advanced the true conception of scientific working. Of course, up to a certain point, it is notorious that the revolt against the purely "subjective methods" in the sixteenth century revived the spirit of inductive research as this had been left by the Greeks; but even with regard to this revolt there are two things which I should like to observe.

In the first place, it seems to me, an altogether disproportionate value has been assigned to Bacon's share in the movement. At most, I think, he deserves to be regarded but as a literary exponent of the Zeitgeist of his century. Himself a philosopher, as distinguished from a man of science, whatever influence his preaching may have had upon the general public, it seems little short of absurd to suppose that it could have produced any considerable effect upon men who were engaged in the practical work of research. And those who read the Novum Organon with a first-hand knowledge of what is required for such research can scarcely fail to agree with his great contemporary Harvey, that he wrote upon science like a Lord Chancellor.

The second thing I should like to observe is, that as the revolt against the purely subjective methods grew in extent and influence it passed to the opposite extreme, which eventually became only less deleterious to the interests of science than was the bondage of authority, and addiction to a priori methods, from which the revolt had set her free. For, without here waiting to trace the history of this matter in detail, I think it ought now to be manifest to everyone who studies it, that up to the commencement of the present century the progress of science in general, and of natural history in particular, was seriously retarded by what may be termed the Bugbear of

Speculation. Fully awakened to the dangers of web-spinning from the ever-fertile resources of their own inner consciousness, naturalists became more and more abandoned to the idea that their science ought to consist in a mere observation of facts, or tabulation of phenomena, without attempt at theorizing upon their philosophical import. If the facts and phenomena presented any such import, that was an affair for men of letters to deal with; but, as men of science, it was their duty to avoid the seductive temptations of the world, the flesh, and the devil, in the form of speculation, deduction, and generalization.

I do not allege that this ideal of natural history was either absolute or universal; but there can be no question that it was both orthodox and general. Even Linneus was express in his limitations of true scientific work in natural history to the collecting and arranging of species of plants and animals. In accordance with this view, the status of a botanist or a zoologist was estimated by the number of specific names, natural habitats, &c., which he could retain in his memory, rather than by any evidences which he might give of intellectual powers in the way of constructive thought. At the most these powers might legitimately exercise themselves only in the direction of taxonomic work; and if a Hales, a Haller, or a Hunter obtained any brilliant results in the way of observation and experiment, their merit was taken to consist in the discovery of facts per se: not in any endeavours they might make in the way of combining their facts under general principles. Even as late in the day as Cuvier this ideal was upheld as the strictly legitimate one for a naturalist to follow; and although Cuvier himself was far from being always loyal to it, he leaves no doubt regarding the estimate in which he held the still greater deviations of his colleagues, St. Hilaire and Lamarck.

Now, these traditional notions touching the severance between the facts of natural history and the philosophy of it, continued more or less to dominate the minds of naturalists until the publication of the Origin of Species, in 1859. Then it was that an epoch was marked in this respect, as in so many other respects where natural history is concerned. For, looking to the enormous results which followed from a deliberate disregard of such traditional canons by Darwin, it has long since become impossible for naturalists, even of the strictest sect, not to perceive that their previous bondage to the law of a mere ritual has been for ever superseded by what verily deserves to be regarded as a new dispensation. Yet it cannot be said, or even so much as suspected, that Darwin's method in any way resembled that of pre-scientific days, the revolt against which led to the straight-laced--and for a long time most salutary--conceptions of method that we have just been noticing. Where, then, is the difference? To me it seems that the difference is as follows; and, if so, that not the least of our many obligations to Darwin as the great organizer of biological science arises from his having clearly displayed the true principle which ought to govern biological research.

To begin with, he nowhere loses sight of the primary distinction between fact and theory; so that, thus far, he loyally follows the spirit of revolt against subjective methods. But, while always holding this distinction clearly in view, his idea of the scientific use of facts is plainly that of furnishing legitimate material for the construction of theories. Natural history is not to him an affair of the herbarium or the cabinet. The collectors and the species-framers are, as it were, his diggers of clay and makers of bricks: even the skilled observers and the trained experimentalists are his mechanics. Valuable as the work of all these men is in itself, its principal value, as he has finally demonstrated, is that which it acquires in rendering possible the work of the architect. Therefore, although he has toiled in all the trades with his own hands, and in each has accomplished some of the best work that has ever been done, the great difference between him and most of his predecessors consists in this,--that while to them the discovery or accumulation of facts was an end, to him it is the means. In their eyes it was enough that the facts should be discovered and recorded. In his eyes the value of facts is due to their power of guiding the mind to a further discovery of principles. And the extraordinary success which attended his work in this respect of generalization immediately brought natural history into line with the other inductive sciences, behind which, in this most important of all respects, she has so seriously fallen. For it was the Origin of Species which first clearly revealed to naturalists as a class, that it was the duty of their science to take as its motto, what is really the motto of natural science in general,

Felix qui potuit rerum cognoscere causas.

Not facts, then, or phenomena, but causes or principles, are the ultimate objects of scientific quest. It remains to ask, How ought this quest to be prosecuted?

Well, in the second place, Darwin has shown that next only to the importance of clearly distinguishing between facts and theories on the one hand, and of clearly recognising the relation between them on the other, is the importance of not being scared by the Bugbear of Speculation. The spirit of speculation is the same as the spirit of science, namely, as we have just seen, a desire to know the causes of things. The hypotheses non fingo of Newton, if taken to mean what it is often understood as meaning, would express precisely the opposite spirit from that in which all scientific research must necessarily take its origin. For if it be causes or principles, as distinguished from facts or phenomena, that constitute the final aim of scientific research, obviously the advancement of such research can be attained only by the framing of hypotheses. And to frame hypotheses is to speculate.

Therefore, the difference between science and speculation is not a difference of spirit; nor, thus far, is it a difference of method. The only difference between them is in the subsequent process of verifying hypotheses. For while speculation, in its purest form, is satisfied to test her explanations only by the degree in which they accord with our subjective ideas of probability--or with the "Illative Sense" of Cardinal Newman,--science is not satisfied to rest in any explanation as final until it shall have been fully verified by an appeal to objective proof. This distinction is now so well and so generally appreciated that I need not dwell upon it. Nor need I wait to go into any details with regard to the so-called canons of verification. My only object is to make perfectly clear, first, that in order to have any question to put to the test of objective verification, science must already have so far employed the method of speculation as to have framed a question to be tested; and, secondly, that the point where science parts company with speculation is the point where this testing process begins.

Now, if these things are so, there can be no doubt that Darwin was following the truest method of inductive research in allowing any amount of latitude to his speculative thought in the direction of scientific theorizing. For it follows from the above distinctions that the danger of speculation does not reside in the width of its range, or even in the impetuosity of its vehemence. Indeed, the wider its reach, and the greater its energy, the better will it be for the interests of science. The only danger of speculation consists in its momentum being apt to carry away the mind from the more laborious work of adequate verification; and therefore a true scientific judgment consists in giving a free rein to speculation on the one hand, while holding ready the break of verification with the other. Now, it is just because Darwin did both these things with so admirable a judgment, that he gave the world of natural history so good a lesson as to the most effectual way of driving the chariot of science.

This lesson we have now all more or less learnt to profit by. Yet no other naturalist has proved himself so proficient in holding the balance true. For the most part, indeed, they have now all ceased to confound the process of speculation per se with the danger of inadequate verification; and therefore the old ideal of natural history as concerned merely with collecting species, classifying affinities, and, in general, tabulating facts, has been well-nigh universally superseded. But this great gain has been attended by some measure of loss. For while not a few naturalists have since erred on the side of insufficiently distinguishing between fully verified principles of evolution and merely speculative deductions therefrom, a still larger number have formed for themselves a Darwinian creed, and regard any further theorizing on the subject of evolution as ipso facto unorthodox.

Having occupied the best years of my life in closely studying the literature of Darwinism, I shall endeavour throughout the following pages to avoid both these extremes. No one in this generation is able to imitate Darwin, either as an observer or a generalizer. But this does not hinder that we should all so far endeavour to follow his method, as always to draw a clear distinction, not merely between observation and deduction, but also between degrees of verification. At all events, my own aim will everywhere be to avoid dogmatism on the one hand, and undue timidity as regards general reasoning on the other. For everything that is said justification will be given; and, as far as prolonged deliberation has enabled me to do so, the exact value of such justification will be rendered by a statement of at least the main grounds on which it rests. The somewhat extensive range of the present treatise, however, will not admit of my rendering more than a small percentage of the facts which in each case go to corroborate the conclusion. But although a great deal must thus be necessarily lost on the one side, I am disposed to think that more will be gained on the other, by presenting, in a terser form than would otherwise be possible, the whole theory of organic evolution as I believe that it will eventually stand. My endeavour, therefore, will be to exhibit the general structure of this theory in what I take to be its strictly logical form, rather than to encumber any of its parts by a lengthy citation of facts. Following this method, I shall in each case give only what I consider the main facts for and against the positions which have to be argued; and in most cases I shall arrange the facts in two divisions, namely, first those of largest generality, and next a few of the most special character that can be found.

As explained in the Preface, the present instalment of the treatise is concerned with the theory of evolution, from the appearance of the Origin of Species in 1859, to the death of its author in 1882; while the second part will be devoted to the sundry post-Darwinian questions which have arisen in the subsequent decade. To the possible criticism that a disproportionate amount of space will thus be allotted to a consideration of these post-Darwinian questions, I may furnish in advance the following reply.

In the first place, besides the works of Darwin himself, there are a number of others which have already and very admirably expounded the evidences, both of organic evolution as a fact, and of natural selection as a cause. Therefore, in the present treatise it seemed needless to go beyond the ground which was covered by my original lectures, namely, a condensed and connected, while at the same time a critical statement of the main evidences, and the main objections, which have thus far been published with reference to the distinctively Darwinian theory. Indeed while re-casting this portion of my lectures for the present publication, I have felt that criticism might be more justly urged from the side of impatience at a reiteration of facts and arguments already so well known. But while endeavouring, as much as possible, to avoid overlapping the previous expositions, I have not carried this attempt to the extent of damaging my own, by omitting any of the more important heads of evidence; and I have sought to invest the latter with some measure of novelty by making good what appears to me a deficiency which has hitherto obtained in the matter of pictorial illustration. In particular, there will be found a tolerably extensive series of woodcuts, serving to represent the more important products of artificial selection. These, like all the other original illustrations, have been drawn either direct from nature or from a comparative study of the best authorities. Nevertheless, I desire it to be understood that the first part of this treatise is intended to retain its original character, as a merely educational exposition of Darwinian teaching-- an exposition, therefore, which, in its present form, may be regarded as a compendium, or hand-book, adapted to the requirements of a general reader, or biological student as distinguished from those of a professed naturalist.

The case, however, is different with the second instalment, which will be published at no very distant date. Here I have not followed with nearly so much closeness the material of my original lectures. On the contrary, I have had in view a special class of readers; and, although I have tried not altogether to sacrifice the more general class, I shall desire it to be understood that I am there appealing to naturalists who are specialists in Darwinism. One must say advisedly, naturalists who are specialists in Darwinism, because, while the literature of Darwinism has become a department of science in itself, there are nowadays many naturalists who, without having paid any close attention to the subject, deem themselves entitled to hold authoritative opinions with regard to it. These men may have done admirable work in other departments of natural history, and yet their opinions on such matters as we shall hereafter have to consider may be destitute of value. As there is no necessary relation between erudition in one department of science and soundness of judgment in another, the mere fact that a man is distinguished as a botanist or zoologist does not in itself qualify him as a critic where specially Darwinian questions are concerned. Thus it happens now, as it happened thirty years ago, that highly distinguished botanists and zoologists prove themselves incapable as judges of general reasoning. It was Darwin's complaint that for many years nearly all his scientific critics either could not, or would not, understand what he had written--and this even as regarded the fundamental principles of his theory, which with the utmost clearness he had over and over again repeated. Now the only difference between such naturalists and their successors of the present day is, that the latter have grown up in a Darwinian environment, and so, as already remarked, have more or less thoughtlessly adopted some form of Darwinian creed. But this scientific creed is not a whit less dogmatic and intolerant than was the more theological one which it has supplanted; and while it usually incorporates the main elements of Darwin's teaching, it still more usually comprises gross perversions of their consequences. All this I shall have occasion more fully to show in subsequent parts of the present work; and allusion is made to the matter here merely for the sake of observing that in future I shall not pay attention to unsupported expressions of opinion from any quarter: I shall consider only such as are accompanied with some statement of the grounds upon which the opinion is held. And, even as thus limited, I do not think it will be found that the following exposition devotes any disproportional amount of attention to the contemporary movements of Darwinian thought, seeing, as we shall see, how active scientific speculation has been in the field of Darwinism since the death of Mr. Darwin.

* * * * *

Leaving, then, these post-Darwinian questions to be dealt with subsequently, I shall now begin a systematic summaryof the evidences in favour of the Darwinian theory, as this was left to the world by Darwin himself.

There is a great distinction to be drawn between the fact of evolution and the manner of it, or between the evidence of evolution as having taken place somehow, and the evidence of the causes which have been concerned in the process. This most important distinction is frequently disregarded by popular writers on Darwinism; and, therefore, in order to mark it as strongly as possible, I will effect a complete separation between the evidence which we have of evolution as a fact, and the evidence which we have as to its method. In other words, not until I shall have fully considered the evidence of organic evolution as a process which somehow or another has taken place, will I proceed to consider how it has taken place, or the causes which Darwin and others have suggested as having probably been concerned in this process.

Confining, then, our attention in the first instance to a proof of evolution considered as a fact, without any reference at all to its method, let us begin by considering the antecedent standing of the matter.

* * * * *

First of all we must clearly recognise that there are only two hypotheses in the field whereby it is possible so much as to suggest an explanation of the origin of species. Either all the species of plants and animals must have been supernaturally created, or else they must have been naturally evolved. There is no third hypothesis possible; for no one can rationally suggest that species have been eternal.

Next, be it observed, that the theory of a continuous transmutation of species is not logically bound to furnish a full explanation of all the natural causes which it may suppose to have been at work. The radical distinction between the two theories consists in the one assuming an immediate action of some supernatural or inscrutable cause, while the other assumes the immediate action of natural--and therefore of possibly discoverable--causes. But in order to sustain this latter assumption, the theory of descent is under no logical necessity to furnish a full proof of all the natural causes which may have been concerned in working out the observed results. We do not know the natural causes of many diseases; but yet no one nowadays thinks of reverting to any hypothesis of a supernatural cause, in order to explain the occurrence of any disease the natural causation of which is obscure. The science of medicine being in so many cases able to explain the occurrence of disease by its hypothesis of natural causes, medical men now feel that they are entitled to assume, on the basis of a wide analogy, and therefore on the basis of a strong antecedent presumption, that all diseases are due to natural causes, whether or not in particular cases such causes happen to have been discovered. And from this position it follows that medical men are not logically bound to entertain any supernatural theory of an obscure disease, merely because as yet they have failed to find a natural theory. And so it is with biologists and their theory of descent. Even if it be fully proved to them that the causes which they have hitherto discovered, or suggested, are inadequate to account for all the facts of organic nature, this would in no wise logically compel them to vacate their theory of evolution, in favour of the theory of creation. All that it would so compel them to do would be to search with yet greater diligence for the natural causes still undiscovered, but in the existence of which they are, by their independent evidence in favour of the theory, bound to believe.

In short, the issue is not between the theory of a supernatural cause and the theory of any one particular natural cause, or set of causes--such as natural selection, use, disuse, and so forth. The issue thus far--or where only the fact of evolution is concerned--is between the theory of a supernatural cause as operating immediately in numberless acts of special creation, and the theory of natural causes as a whole, whether these happen, or do not happen, to have been hitherto discovered.

This much by way of preliminaries being understood, we have next to notice that whichever of the two rival theories we choose to entertain, we are not here concerned with any question touching the origin of life. We are concerned only with the origin of particular forms of life--that is to say, with the origin of species. The theory of descent starts from life as a datum already granted. How life itself came to be, the theory of descent, as such, is not concerned to show. Therefore, in the present discussion, I will take the existence of life as a fact which does not fall within the range of our present discussion. No doubt the question as to the origin of life is in itself a deeply interesting question, and although in the opinion of most biologists it is a question which we may well hope will some day fall within the range of science to answer, at present, it must be confessed, science is not in a position to furnish so much as any suggestion upon the subject; and therefore our wisdom as men of science is frankly to acknowledge that such is the case.

* * * * *

We are now in a position to observe that the theory of organic evolution is strongly recommended to our acceptance on merely antecedent grounds, by the fact that it is in full accordance with what is known as the principle of continuity. By the principle of continuity is meant the uniformity of nature, in virtue of which the many and varied processes going on in nature are due to the same kind of method, i. e. the method of natural causation. This conception of the uniformity of nature is one that has only been arrived at step by step through a long and arduous course of human experience in the explanation of natural phenomena. The explanations of such phenomena which are first given are always of the supernatural kind; it is not until investigation has revealed the natural causes which are concerned that the hypotheses of superstition give way to those of science. Thus it follows that the hypotheses of superstition which are the latest in yielding to the explanations of science, are those which refer to the more recondite cases of natural causation; for here it is that methodical investigation is longest in discovering the natural causes. Thus it is only by degrees that fetishism is superseded by what now appears a common-sense interpretation of physical phenomena; that exorcism gives place to medicine; alchemy to chemistry; astrology to astronomy; and so forth. Everywhere the miraculous is progressively banished from the field of explanation by the advance of scientific discovery; and the places where it is left longest in occupation are those where the natural causes are most intricate or obscure, and thus present the greatest difficulty to the advancing explanations of science. Now, in our own day there are but very few of these strongholds of the miraculous left. Nearly the whole field of explanation is occupied by naturalism, so that no one ever thinks of resorting to supernaturalism except in the comparatively few cases where science has not yet been able to explore the most obscure regions of causation. One of these cases is the origin of life; and, until quite recently, another of these cases was the origin of species. But now that a very reasonable explanation of the origin of species has been offered by science, it is but in accordance with all previous historical analogies that many minds should prove themselves unable all at once to adjust themselves to the new ideas, and thus still linger about the more venerable ideas of supernaturalism. But we are now in possession of so many of these historical analogies, that all minds with any instincts of science in their composition have grown to distrust, on merely antecedent grounds, any explanation which embodies a miraculous element. Such minds have grown to regard all these explanations as mere expressions of our own ignorance of natural causation; or, in other words, they have come to regard it as an a priori truth that nature is everywhere uniform in respect of method or causation; that the reign of law universal; the principle of continuity ubiquitous.

Now, it must be obvious to any mind which has adopted this attitude of thought, that the scientific theory of natural descent is recommended by an overwhelming weight of antecedent presumption, as against the dogmatic theory of supernatural design.

To begin with, we must remember that the fact of evolution--or, which is the same thing, the fact of continuity in natural causation--has now been unquestionably proved in so many other and analogous departments of nature, that to suppose any interruption of this method as between species and species becomes, on grounds of such analogy alone, well-nigh incredible. For example, it is now a matter of demonstrated fact that throughout the range of inorganic nature the principles of evolution have obtained. It is no longer possible for any one to believe with our forefathers that the earth's surface has always existed as it now exists. For the science of geology has proved to demonstration that seas and lands are perpetually undergoing gradual changes of relative positions--continents and oceans supplanting each other in the course of ages, mountain-chains being slowly uplifted, again as slowly denuded, and so forth. Moreover, and as a closer analogy, within the limits of animate nature we know it is the universal law that every individual life undergoes a process of gradual development; and that breeds, races, or strains, may be brought into existence by the intentional use of natural processes--the results bearing an unmistakeable resemblance to what we know as natural species. Again, even in the case of natural species themselves, there are two considerations which present enormous force from an antecedent point of view. The first is that organic forms are only then recognised as species when intermediate forms are absent. If the intermediate forms are actually living, or admit of being found in the fossil state, naturalists forthwith regard the whole series as varieties, and name all the members of it as belonging to the same species. Consequently it becomes obvious that naturalists, in their work of naming species, may only have been marking out the cases where intermediate or connecting forms have been lost to observation. For example, here we have a diagram representing a very unusually complete series of fossil shells, which within the last few years has been unearthed from the Tertiary lake basins of Slavonia. Before the series was completed, some six or eight of the then disconnected forms were described as distinct species; but as soon as the connecting forms were found--showing a progressive modification from the older to the newer beds,--the whole were included as varieties of one species.

Of course, other cases of the same kind might be adduced, and therefore, as just remarked, in their work of naming species naturalists may only have been marking out the cases where intermediate forms have been lost to observation. And this possibility becomes little less than a certainty when we note the next consideration which I have to adduce, namely, that in all their systematic divisions of plants and animals in groups higher than species--such as genera, families, orders, and the rest--naturalists have at all times recognised the fact that the one shades off into the other by such imperceptible gradations, that it is impossible to regard such divisions as other than conventional. It is important to remember that this fact was fully recognised before the days of Darwin. In those days the scientifically orthodox doctrine was, that although species were to be regarded as fixed units, bearing the stamp of a special creation, all the higher taxonomic divisions were to be considered as what may be termed the artificial creation of naturalists themselves. In other words, it was believed, and in many cases known, that if we could go far enough back in the history of the earth, we should everywhere find a tendency to mutual approximation between allied groups of species; so that, for instance, birds and reptiles would be found to be drawing nearer and nearer together, until eventually they would seem to become fused in a single type; that the existing distinctions between herbivorous and carnivorous mammals would be found to do likewise; and so on with all the larger group-distinctions, at any rate within the limits of the same sub-kingdoms. But although naturalists recognised this even in the pre-Darwinian days, they stoutly believed that a great exception was to be made in the case of species. These, the lowest or initial members of their taxonomic series, they supposed to be permanent--the miraculously created units of organic nature. Now, all that I have at present to remark is, that this pre-Darwinian exception which was made in favour of species to the otherwise recognised principle of gradual change, was an exception which can at no time have been recommended by any antecedent considerations. At all times it stood out of analogy with the principle of continuity; and, as we shall fully find in subsequent chapters, it is now directly contradicted by all the facts of biological science.

There remains one other fact of high generality to which prominent attention should be drawn from the present, or merely antecedent, point of view. On the theory of special creation no reason can be assigned why distinct specific types should present any correlation, either in time or in space, with their nearest allies; for there is evidently no conceivable reason why any given species, A, should have been specially created on the same area and at about the same time as its nearest representative, B,--still less, of course, that such should be a general rule throughout all the thousands and millions of species which have ever inhabited the earth. But, equally of course, on the theory of a natural evolution this is so necessary a consequence, that if no correlation of such a two-fold kind were observable, the theory would be negatived. Thus the question whether there be any indication of such a two-fold correlation may be regarded as a test-question as between the two theories; for although the vast majority of extinct species have been lost to science, there are a countless number of existing species which furnish ample material for answering the question. And the answer is so unequivocal that Mr. Wallace, who is one of our greatest authorities on geographical distribution, has laid it down as a general law, applicable to all the departments of organic nature, that, so far as observation can extend, "every species has come into existence coincident both in space and time with a pre-existing and closely allied species." As it appears to me that the significance of these words cannot be increased by any comment upon them, I will here bring this introductory chapter to a close.

CHAPTER II.

CLASSIFICATION.

The first line of direct evidence in favour of organic evolution which I shall open is that which may be termed the argument from Classification.

It is a matter of observable fact that different forms of plants and animals present among themselves more or less pronounced resemblances. From the earliest times, therefore, it has been the aim of philosophical naturalists to classify plants and animals in accordance with these resemblances. Of course the earliest attempts at such classification were extremely crude. The oldest of these attempts with which we are acquainted--namely, that which is presented in the books of Genesis and Leviticus--arranges the whole vegetable kingdom in three simple divisions of Grass, Herbs, and Trees; while the animal kingdom is arranged with almost equal simplicity with reference, first to habitats in water, earth, or air, and next as to modes of progression. These, of course, were what may be termed common-sense classifications, having reference merely to external appearances and habits of life. But when Aristotle laboriously investigated the comparative anatomy of animals, he could not fail to perceive that their entire structures had to be taken into account in order to classify them scientifically; and, also, that for this purpose the internal parts were of quite as much importance as the external. Indeed, he perceived that they were of greatly more importance in this respect, inasmuch as they presented so many more points for comparison; and, in the result, he furnished an astonishingly comprehensive, as well as an astonishingly accurate classification of the larger groups of the animal kingdom. On the other hand, classification of the vegetable kingdom continued pretty much as it had been left by the book of Genesis--all plants being divided into three groups, Herbs, Shrubs, and Trees. Nor was this primitive state of matters improved upon till the sixteenth century, when Gesner (1516-1565), and still more Calpino (1519-1603), laid the foundations of systematic botany.

But the more that naturalists prosecuted their studies on the anatomy of plants and animals, the more enormously complex did they find the problem of classification become. Therefore they began by forming what are called artificial systems, in contradistinction to natural systems. An artificial system of classification is a system based on the more or less arbitrary selection of some one part, or set of parts; while a natural classification is one that is based upon a complete knowledge of all the structures of all the organisms which are classified.

Thus, the object of classification has been that of arranging organisms in accordance with their natural affinities, by comparing organism with organism, for the purpose of ascertaining which of the constituent organs are of the most invariable occurrence, and therefore of the most typical signification. A porpoise, for instance, has a large number of teeth, and in this feature resembles most fish, while it differs from all mammals. But it also gives suck to its young. Now, looking to these two features alone, should we say that a porpoise ought to be classed as a fish or as a mammal? Assuredly as a mammal; because the number of teeth is a very variable feature both in fish and mammals, whereas the giving of suck is an invariable feature among mammals, and occurs nowhere else in the animal kingdom. This, of course, is chosen as a very simple illustration. Were all cases as obvious, there would be but little distinction between natural and artificial systems of classification. But it is because the lines of natural affinity are, as it were, so interwoven throughout the organic world, and because there is, in consequence, so much difficulty in following them, that artificial systems have to be made in the first instance as feelers towards eventual discovery of the natural system. In other words, while forming their artificial systems of classification, it has always been the aim of naturalists--whether consciously or unconsciously--to admit as the bases of their systems those characters which, in the then state of their knowledge, seemed most calculated to play an important part in the eventual construction of the natural system. If we were dealing with the history of classification, it would here be interesting to note how the course of it has been marked by gradual change in the principles which naturalists adopted as guides to the selection of characters on which to found their attempts at a natural classification. Some of these changes, indeed, I shall have to mention later on; but at present what has to be specially noted is, that through all these changes of theory or principle, and through all the ever-advancing construction of their taxonomic science, naturalists themselves were unable to give any intelligible reason for the faith that was in them--or the faith that over and above the artificial classifications which were made for the mere purpose of cataloguing the living library of organic nature, there was deeply hidden in nature itself a truly natural classification, for the eventual discovery of which artificial systems might prove to be of more or less assistance.

Linneus, for example, expressly says--"You ask me for the characters of the natural orders; I confess that I cannot give them." Yet he maintains that, although he cannot define the characters, he knows, by a sort of naturalist's instinct, what in a general way will subsequently be found to be the organs of most importance in the eventual grouping of plants under a natural system. "I will not give my reasons for the distribution of the natural orders which I have published," he said: "you, or some other person, after twenty or after fifty years, will discover them, and see that I was right."

Thus we perceive that in forming their provisional or artificial classifications, naturalists have been guided by an instinctive belief in some general principle of natural affinity, the character of which they have not been able to define; and that the structures which they selected as the bases of their classifications when these were consciously artificial, were selected because it seemed that they were the structures most likely to prove of use in subsequent attempts at working out the natural system.

This general principle of natural affinity, of which all naturalists have seen more or less well-marked evidence in organic nature, and after which they have all been feeling, has sometimes been regarded as natural, but more often as supernatural. Those who regarded it as supernatural took it to consist in a divine ideal of creation according to types, so that the structural affinities of organisms were to them expressions of an archetypal plan, which might be revealed in its entirety when all organisms on the face of the earth should have been examined. Those, on the other hand, who regarded the general principle of affinity as depending on some natural causes, for the most part concluded that these must have been utilitarian causes; or, in other words, that the fundamental affinities of structure must have depended upon fundamental requirements of function. According to this view, the natural classification would eventually be found to stand upon a basis of physiology. Therefore all the systems of classification up to the earlier part of the present century went upon the apparent axiom, that characters which are of most importance to the organisms presenting them must be characters most indicative of natural affinities. But the truth of the matter was eventually found to be otherwise. For it was eventually found that there is absolutely no correlation between these two things; that, therefore, it is a mere chance whether or not organs which are of importance to organisms are likewise of importance as guides to classification; and, in point of fact, that the general tendency in this matter is towards an inverse instead of a direct proportion. More often than not, the greater the value of a structure for the purpose of indicating natural affinities, the less is its value to the creatures presenting it.

Enough has now been said to show three things. First, that long before the theory of descent was entertained by naturalists, naturalists perceived the fact of natural affinities, and did their best to construct a natural system of classification for the purpose of expressing such affinities. Second, that naturalists had a kind of instinctive belief in some one principle running through the whole organic world, which thus served to bind together organisms in groups subordinate to groups--that is, into species, genera, orders, families, classes, sub-kingdoms, and kingdoms. Third, that they were not able to give any very intelligible reason for this faith that was in them; sometimes supposing the principle in question to be that of a supernatural plan of organization, sometimes regarding it as dependent on conditions of physiology, and sometimes not attempting to account for it at all.

Of course it is obvious that the theory of descent furnishes the explanation which is required. For it is now evident to evolutionists, that although these older naturalists did not know what they were doing when they were tracing these lines of natural affinity, and thus helping to construct a natural classification--I say it is now evident to evolutionists that these naturalists were simply tracing the lines of genetic relationship. The great principle pervading organic nature, which was seen so mysteriously to bind the whole creation together as in a nexus of organic affinity, is now easily understood as nothing more or less than the principle of Heredity. Let us, therefore, look a little more closely at the character of this network, in order to see how far it lends itself to this new interpretation.

The first thing that we have to observe about the nexus is, that it is a nexus-- not a single line, or even a series of parallel lines. In other words, some time before the theory of descent was seriously entertained, naturalists for the most part had fully recognised that it was impossible to arrange either plants or animals, with respect to their mutual affinities, in a ladder-like series (as was supposed to be the type of classification by the earlier systematists), or even in map-like groups (as was supposed to be the type by Linneus). And similarly, also, with respect to grades of organization. In the case of the larger groups, indeed, it is usually possible to say that the members of this group as a whole are more highly organized than the members of that group as a whole; so that, for instance, we have no hesitation in regarding the Vertebrata as more highly organized than the Invertebrata, Birds than Reptiles, and so on. But when we proceed to smaller subdivisions, such as genera and species, it is usually impossible to say that the one type is more highly organized than another type. A horse, for instance, cannot be said to be more highly organized than a zebra or an ass; although the entire horse-genus is clearly a more highly organized type than any genus of animal which is not a mammal.

In view of these facts, therefore, the system of classification which was eventually arrived at before the days of Darwin, was the system which naturalists likened to a tree; and this is the system which all naturalists now agreed upon as the true one. According to this system, a short trunk may be taken to represent the lowest organisms which cannot properly be termed either plants or animals. This short trunk soon separates into two large trunks, one of which represents the vegetable and the other the animal kingdom. Each of these trunks then gives off large branches signifying classes, and these give off smaller, but more numerous branches, signifying families, which ramify again into orders, genera, and finally into the leaves, which may be taken to represent species. Now, in such a representative tree of life, the height of any branch from the ground may be taken to indicate the grade of organization which the leaves, or species, present; so that, if we picture to ourselves such a tree, we may understand that while there is a general advance of organization from below upwards, there are many deviations in this respect. Sometimes leaves growing on the same branch are growing at a different level--especially, of course, if the branch be a large one, corresponding to a class or sub-kingdom. And sometimes leaves growing on different branches are growing at the same level: that is to say, although they represent species belonging to widely divergent families, orders, or even classes, it cannot be said that the one species is more highly organized than the other.

Now, this tree-like arrangement of species in nature is an arrangement for which Darwin is not responsible. For, as we have seen, the detecting of it has been due to the progressive work of naturalists for centuries past; and even when it was detected, at about the commencement of the present century, naturalists were confessedly unable to explain the reason of it, or what was the underlying principle that they were engaged in tracing when they proceeded ever more and more accurately to define these ramifications of natural affinity. But now, as just remarked, we can clearly perceive that this underlying principle was none other than Heredity as expressed in family likeness,--likeness, therefore, growing progressively more unlike with remoteness of ancestral relationship. For thus only can we obtain any explanation of the sundry puzzles and apparent paradoxes, which a working out of their natural classifications revealed to botanists and zoologists during the first half of the present century. It will now be my endeavour to show how these puzzles and paradoxes are all explained by the theory that natural affinities are merely the expression of genetic affinities.

First of all, and from the most general point of view, it is obvious that the tree-like system of classification, which Darwin found already and empirically worked out by the labours of his predecessors, is as suggestive as anything could well be of the fact of genetic relationship. For this is the form that every tabulation of family pedigree must assume; and therefore the mere fact that a scientific tabulation of natural affinities was eventually found to take the form of a tree, is in itself highly suggestive of the inference that such a tabulation represents a family tree. If all species were separately created, there can be no assignable reason why the ideas of earlier naturalists touching the form which a natural classification would eventually assume should not have represented the truth--why, for example, it should not have assumed the form of a ladder (as was anticipated in the seventeenth century), or of a map (as was anticipated in the eighteenth), or, again, of a number of wholly unrelated lines, circles, &c. (as certain speculative writers of the present century have imagined). But, on the other hand, if all species were separately and independently created, it becomes virtually incredible that we should everywhere observe this progressive arborescence of characters common to larger groups into more and more numerous, and more and more delicate, ramifications of characters distinctive only of smaller and smaller groups. A man would be deemed insane if he were to attribute the origin of every branch and every twig of a real tree to a separate act of special creation;

and although we have not been able to witness the growth of what we may term in a new sense the Tree of Life, the structural relations which are now apparent between its innumerable ramifications bear quite as strong a testimony to the fact of their having been due to an organic growth, as is the testimony furnished by the branches of an actual tree.

Or, to take another illustration. Classification of organic forms, as Darwin, Lyell, and H鋑kel have pointed out, strongly resembles the classification of languages. In the case of languages, as in the case of species, we have genetic affinities strongly marked; so that it is possible to some extent to construct a Language-tree, the branches of which shall indicate, in a diagrammatic form, the progressive divergence of a large group of languages from a common stock. For instance, Latin may be regarded as a fossil language, which has given rise to a group of living languages--Italian, Spanish, French, and, to a large extent, English. Now what would be thought of a philologist who should maintain that English, French, Spanish, and Italian were all specially created languages--or languages separately constructed by the Deity, and by as many separate acts of inspiration communicated to the nations which now speak them--and that their resemblance to the fossil form, Latin, must be attributed to special design? Yet the evidence of the natural transmutation of species is in one respect much stronger than that of the natural transmutation of languages--in respect, namely, of there being a vastly greater number of cases all bearing testimony to the fact of genetic relationship.

But, quitting now this most general point of view--or the suggestive fact that what we have before us is a tree--let us next approach this tree for the purpose of examining its structure more in detail. When we do this, the fact of next greatest generality which we find is as follows.

In cases where a very old form of life has continued to exist unmodified, so that by investigation of its anatomy we are brought back to a more primitive type of structure than that of the newer forms growing higher up upon the same branch, two things are observable. In the first place, the old form is less differentiated than the newer ones; and, in the next place, it is seen much more closely to resemble types of structure belonging to some of the other and larger branches of the tree. The organization of the older form is not only simpler; but it is, as naturalists say, more generalized. It comprises within itself characters belonging to its own branch, and also characters belonging to neighbouring branches, or to the trunk from which allied branches spring. Hence it becomes a general rule of classification, that it is by the lowest, or by the oldest, forms of any two natural groups that the affinities between the two groups admit of being best detected. And it is obvious that this is just what ought to be the case on the theory of descent with divergent modification; while, upon the alternative theory of special creation, no reason can be assigned why the lowest or the oldest types should thus combine the characters which afterwards become severally distinctive of higher or newer types.

Again, I have already alluded to the remarkable fact that there is no correlation between the value of structures to the organisms which present them, and their value to the naturalist for the purpose of tracing natural affinity; and I have remarked that up to the close of the last century it was regarded as an axiom of taxonomic science, that structures which are of most importance to the animals or plants possessing them must likewise prove of most importance in any natural system of classification. On this account, all attempts to discover the natural classification went upon the supposition that such a direct proportion must obtain--with the result that organs of most physiological importance were chosen as the bases of systematic work. And when, in the earlier part of the present century, De Candolle found that instead of a direct there was usually an inverse proportion between the functional and the taxonomic value of a structure, he was unable to suggest any reason for this apparently paradoxical fact. For, upon the theory of special creation, no reason can be assigned why organs of least importance to organisms should prove of most importance as marks of natural affinity. But on the theory of descent with progressive modification the apparent paradox is at once explained. For it is evident that organs of functional importance are, other things equal, the organs which are most likely to undergo different modifications in different lines of family descent, and therefore in time to have their genetic relationships in these different lines obscured. On the other hand, organs or structures which are of no functional importance are never called upon to change in response to any change of habit, or to any change in the conditions of life. They may, therefore, continue to be inherited through many different lines of family descent, and thus afford evidence of genetic relationship where such evidence fails to be given by any of the structures of vital importance, which in the course of many generations have been required to change in many ways according to the varied experiences of different branches of the same family. Here, then, we have an empirically discovered rule in the science of classification, the raison d'être of which we are at once able to appreciate upon the theory of evolution, whereas no possible explanation of why it should ever have become a rule could be furnished upon the theory of special creation.

Here, again, is another empirically determined rule. The larger the number, as distinguished from the importance, of structures which are found common to different groups, the greater becomes their value as guides to the determination of natural affinity. Or, as Darwin puts it, "the value of an aggregate of characters, even when none are important, alone explains the aphorism enunciated by Linneus, namely, that the characters do not give the genus, but the genus gives the characters; for this seems founded on the appreciation of many trifling points of resemblance, too slight to be defined[1]."

[1] Origin of Species, p. 367.

Now it is evident, without comment, of how much value aggregates of characters ought to be in classification, if the ultimate meaning of classification be that of tracing lines of pedigree; whereas, if this ultimate meaning were that of tracing divine ideals manifested in special creation, we can see no reason why single characters are not such sure tokens of a natural arrangement as are aggregates of characters, even though the latter be in every other respect unimportant. For, on the special creation theory, we cannot explain why an assemblage, say of four or five trifling characters, should have been chosen to mark some unity of plan, rather than some one character of functional importance, which would have served at least equally well any such hypothetical purpose. On the other hand, as Darwin remarks, "we care not how trifling a character may be--let it be the mere inflection of the angle of the jaw, the manner in which an insect's wing is folded, whether the skin be covered with hair or feathers--if it prevail throughout many and different species, especially those having very different habits of life, it assumes high value; for we can account for its presence in so many forms, with such different habits, only by inheritance from a common parent. We may err in this respect in regard to single points of structure, but when several characters, let them be ever so trifling, concur throughout a large group of beings having different habits, we may feel almost sure, on the theory of descent, that these characters have been inherited from a common ancestor; and we know that such aggregated characters have especial value in classification[2]."

[2] Origin of Species, p. 372.

It is true that even a single character, if found common to a large number of forms, while uniformly absent from others, is also regarded by naturalists as of importance for purposes of classification, although they recognise it as of a value subordinate to that of aggregates of characters. But this also is what we should expect on the theory of descent. If even any one structure be found to run through a number of animals presenting different habits of life, the readiest explanation of the fact is to be found in the theory of descent; but this does not hinder that if several such characters always occur together, the inference of genetic relationship is correspondingly confirmed. And the fact that before this inference was ever drawn, naturalists recognised the value of single characters in proportion to their constancy, and the yet higher value of aggregates of characters in proportion to their number--this fact shows that in their work of classification naturalists empirically observed the effects of a cause which we have now discovered, to wit, hereditary transmission of characters through ever-widening groups of changing species.

There is another argument which appears to tell strongly in favour of the theory of descent. We have just seen that non-adaptive structures, not being required to change in response to change of habits or conditions of life, are allowed to persist unchanged through many generations, and thus furnish exceptionally good guides in the science of classification--or, according to our theory, in the work of tracing lines of pedigree. But now, the converse of this statement holds equally true. For it often happens that adaptive structures are required to change in different lines of descent in analogous ways, in order to meet analogous needs; and, when such is the case, the structures concerned have to assume more or less close resemblances to one another, even though they have severally descended from quite different ancestors. The paddles of a whale, for instance, most strikingly resemble the fins of a fish as to their outward form and movements; yet, on the theory of descent, they must be held to have had a widely different parentage. Now, in all such cases where there is thus what is called an analogous (or adaptive) resemblance, as distinguished from what is called an homologous (or anatomical) resemblance--in all such cases it is observable that the similarities do not extend further into the structure of the parts than it is necessary that they should extend, in order that the structures should both perform the same functions. The whole anatomy of the paddles of a whale is quite unlike that of the fins of a fish--being, in fact, that of the fore-limb of a mammal. The change, therefore, which the fore-limb has here undergone to suit it to the aquatic habits of this mammal, is no greater than was required for that purpose: the change has not extended to any one feature of anatomical significance. This, of course, is what we should expect on the theory of descent with modification of ancestral characters; but on the theory of special creation it is not intelligible why there should always be so marked a distinction between resemblances as analogical or adaptive, and resemblances as homological or of meaning in reference to a natural classification. To take another and more detailed instance, the Tasmanian wolf is an animal separated from true wolves in a natural system of classification. Yet its jaws and teeth bear a strong general resemblance to those of all the dog tribe, although there are differences of anatomical detail. In particular, while the dogs all have on each side of the upper jaw four pre-molars and two molars, the Tasmanian wolf has three pre-molars and four molars. Now there is no reason, so far as their common function of dealing with flesh is concerned, why the teeth of the Tasmanian wolf should not have resembled homologically as well as analogically the teeth of a true wolf; and therefore we cannot assign any intelligible reason why, if all the species of the dog genus were separately created with one pattern of teeth, the unallied Tasmanian wolf should have been furnished with what is practically the same pattern from a functional point of view, while differing from a structural point of view. But, of course, on the theory of descent with modification, we can well understand why similarities of habit should have led to similarities of structural appearance of an adaptive kind in different lines of descent, without there being any trace of such real or anatomical similarities as could possibly point to genetic relationship.

Lastly, to adduce the only remaining argument from classification which I regard as of any considerable weight, naturalists have found it necessary, while constructing their natural classifications, to set great store on what Mr. Darwin calls "chains of affinities." Thus, for instance, "nothing can be easier than to define a number of characters common to all birds; but with crustaceans any such definition has hitherto been found impossible. There are crustaceans at the opposite ends of the series, which have hardly a character in common; yet the species at both ends, from being plainly allied to others, and these to others, and so onwards, can be recognised as unequivocally belonging to this, and to no other class of the articulata[3]." Now it is evident that this progressive modification of specific types--where it cannot be said that the continuity of resemblance is anywhere broken, and yet terminates in modification so great that but for the connecting links no one could divine a natural relationship between the extreme members of the series,--it is evident that such chains of affinity speak most strongly in favour of a transmutation of the species concerned, while it is impossible to suggest any explanation of the fact in terms of the rival theory. For if all the links of such a chain were separately forged by as many acts of special creation, we can see no reason why B should resemble A, C resemble B, and so on, but with ever slight though accumulating differences, until there is no resemblance at all between A and Z.

[3] Origin of Species, pp. 368-9.

* * * * *

I hope enough has now been said to show that all the general principles and particular facts appertaining to the natural classification of plants and animals, are precisely what they ought to be according to the theory of genetic descent; while no one of them is such as might be--and, indeed, used to be-- expected upon the theory of special creation. Therefore, the only possible way in which all this uniform body of direct evidence can be met by a supporter of the latter theory, is by falling back upon the argument from ignorance. We do not know, it may be said, what hidden reasons there may have been for following all these general principles in the separate creation of specific types. Now, it is evident that this is a form of argument which admits of being brought against all the actual--and even all the possible--lines of evidence in favour of evolution. Therefore I deem it desirable thus early in our proceedings to place this argument from ignorance on its proper logical footing.

If there were any independent evidence in favour of special creation as a fact, then indeed the argument from ignorance might be fairly used against any sceptical cavils regarding the method. In this way, for example, Bishop

Butler made a legitimate use of the argument from ignorance when he urged that it is no reasonable objection against a revelation, otherwise accredited, to show that it has been rendered in a form, or after a method, which we should not have antecedently expected. But he could not have legitimately employed this argument, except on the supposition that he had some independent evidence in favour of the revelation; for, in the absence of any such independent evidence, appeal to the argument from ignorance would have become a mere begging of the question, by simply assuming that a revelation had been made. And thus it is in the present case. A man, of course, may quite legitimately say, Assuming that the theory of special creation is true, it is not for us to anticipate the form or method of the process. But where the question is as to whether or not the theory is true, it becomes a mere begging of this question to take refuge in the argument from ignorance, or to represent in effect that there is no question to be discussed. And if, when the form or method is investigated, it be found everywhere charged with evidence in favour of the theory of descent, the case becomes the same as that of a supposed revelation, which has been discredited by finding that all available evidence points to a natural growth. In short, the argument from ignorance is in any case available only as a negative foil against destructive criticism: in no case has it any positive value, or value of a constructive kind. Therefore, if a theory on any subject is destitute of positive evidence, while some alternative theory is in possession of such evidence, the argument from ignorance can be of no logical use to the former, even though it maybe of such use to the latter. For it is only the possession of positive evidence which can furnish a logical justification of the argument from ignorance: in the absence of such evidence, even the negative value of the argument disappears, and it then implies nothing more than the gratuitous assumption of a theory.

* * * * *

I will now sum up the various considerations which have occupied us during the present chapter.

First of all we must take note that the classification of plants and animals in groups subordinate to groups is not merely arbitrary, or undertaken only for a matter of convenience and nomenclature--such, for instance, as the classification of stars in constellations. On the contrary, the classification of a naturalist differs from that of an astronomer, in that the objects which he has to classify present structural resemblances and structural differences in numberless degrees; and it is the object of his classification to present a tabular statement of these facts. Now, long before the theory of evolution was entertained, naturalists became fully aware that these facts of structural resemblances running through groups subordinate to groups were really facts of nature, and not merely poetic imaginations of the mind. No one could dissect a number of fishes without perceiving that they were all constructed on one anatomical pattern, which differed considerably from the equally uniform pattern on which all mammals were constructed, even although some mammals bore an extraordinary resemblance to fish in external form and habits of life. And similarly with all the smaller divisions of the animal and vegetable kingdoms. Everywhere investigation revealed the bonds of close structural resemblances between species of the same genus, resemblance less close between genera of the same family, resemblance still less close between families of the same order, resemblance yet more remote between orders of the same class, and resemblance only in fundamental features between classes of the same sub-kingdom, beyond which limit all anatomical resemblance was found to disappear--the different sub-kingdoms being formed on wholly different patterns. Furthermore, in tracing all these grades of structural relationship, naturalists were slowly led to recognise that the form which a natural classification must eventually assume would be that of a tree, wherein the constituent branches would display a progressive advance of organization from below upwards.

Now we have seen that although this tree-like arrangement of natural groups was as suggestive as anything could well be of all the forms o?life being bound together by the ties of genetic relationship, such was not the inference which was drawn from it. Dominated by the theory of special creation, naturalists either regarded the resemblance of type subordinate to type as expressive of divine ideals manifested in such creation, or else contented themselves with investigating the facts without venturing to speculate upon their philosophical import. But even those naturalists who abstained from committing themselves to any theory of archetypal plans, did not doubt that facts so innumerable and so universal must have been due to some one co-ordinating principle--that, even though they were not able to suggest what it was, there must have been some hidden bond of connexion running through the whole of organic nature. Now, as we have seen, it is manifest to evolutionists that this hidden bond can be nothing else than heredity; and, therefore, that these earlier naturalists, although they did not know what they were doing, were really tracing the lines of genetic descent as revealed by degrees of structural resemblance,--that the arborescent grouping of organic forms which their labours led them to begin, and in large measure to execute, was in fact a family tree of life.

Here, then, is the substance of the argument from classification. The mere fact that all organic nature thus incontestably lends itself to a natural arrangement of group subordinate to group, when due regard is paid to degrees of anatomical resemblance--this mere fact of itself tells so weightily in favour of descent with progressive modification in different lines, that even if it stood alone it would be entitled to rank as one of our strongest pieces of evidence. But, as we have seen, it does not stand alone. When we look beyond this large and general fact of all the innumerable forms of life being thus united in a tree-like system by an unquestionable relationship of some kind, to those smaller details in the science of classification which have been found most useful as guides for this kind of research, then we find that all these details, or empirically discovered rules, are exactly what we should have expected them to be, supposing the real meaning of classification to have been that of tracing lines of pedigree.

In particular, we have seen that the most archaic types are both simpler in their organization and more generalized in their characters than are the more recent types--a fact of which no explanation can be given on the theory of special creation. But, upon the theory of natural evolution, we can without difficulty understand why the earlier forms should have been the simpler forms, and also why they should have been the most generalized. For it is out of the older forms that the newer must have grown; and, as they multiplied, they must have become more and more differentiated.

Again, we have seen that there is no correlation between the importance of any structure from a classificatory point of view, and the importance of that structure to the organism which presents it. On the contrary, it is a general rule that "the less any part of the organization is concerned with special habits, the more important it becomes for classification." Now, from the point of view of special creation it is unintelligible why unity of ideal should be most manifested by least important structures, whereas from the point of view of evolution it is to be expected that these life-serving structures should have been most liable to divergent modification in divergent lines of descent, or in adaptation to different conditions of life, while the trivial or less important characters should have been allowed to remain unmodified. Thus we can now understand why all primitive classifications were wrong in principle when they went upon the assumption that divine ideals were best exhibited by resemblances between life-serving (and therefore adaptive) structures, with the result that whales were classed with fishes, birds with bats, and so on. Nevertheless, these primitive naturalists were quite logical; for, from the premises furnished by the theory of special creation, it is much more reasonable to expect that unity of ideal should be shown in plainly adaptive characters than in trivial and more or less hidden anatomical characters. Moreover, long after biological science had ceased consciously to follow any theological theory, the apparent axiom continued to be entertained, that structures of most importance to organisms must also be structures of most importance to systematists. And when at last, in the present century, this was found not to be the case, no reason could be suggested why it was not the case. But now we are able fully to explain this apparent anomaly.

Once more, we have seen that aggregates of characters presenting resemblances to one another have always been found to be of special importance as guides to classification. This, of course, is what we should have expected, if the real meaning of classification be that of tracing lines of pedigree; but on the theory of special creation no reason can be assigned why single characters are not such sure tokens of a natural arrangement as are aggregates of characters, however trivial the latter may be. For it is obvious that unity of ideal might have been even better displayed by everywhere maintaining the pattern of some one important structure, than by doing so in the case of several unimportant structures. Take an analogous instance from human contrivances. Unity of ideal in the case of gun-making would be shown by the same principles of mechanism running through all the different sizes and shapes of gun-locks, rather than by the ornamental patterns engraved upon the outside. Yet it must be supposed that in the mechanisms assumed to have been constructed by special creation, it was the trivial details rather than the fundamental principles of these mechanisms which were chosen by the Divinity to display his ideals.

And this leads us to the next consideration--namely, that when in two different lines of descent animals happen to adopt similar habits of life, the modifications which they undergo in order to fit them for these habits often induces striking resemblances of structure between the two animals, as in the case of whales and fish. But in all such instances it is invariably found that the resemblance is only superficial and apparent: not anatomical or real. In other words, the resemblance does not extend further than it is necessary that it should, if both sets of organs are to be adapted to perform the same functions. Now this, again, is just what one would expect to find as the universal rule on the theory of descent, with modification of ancestral characters. But, on the opposite theory of special creation, I know not how it is to be explained that among so many instances of close superficial resemblance between creatures belonging to different branches of the tree of life, there are no instances of any real or anatomical resemblance. So far as their structures are adapted to perform a common function, there is in all such cases what may be termed a deceptive appearance of some unity of ideal; but, when carefully examined, it is always found that two apparently identical structures occurring on different branches of the classificatory tree are in fact fundamentally different in respect of their structural plan.

Lastly, we have seen that one of the guiding principles of classification has been empirically found to consist in setting a high value on "chains of affinities." That is to say, naturalists not unfrequently meet with a long series of progressive modifications of type, which, although it cannot be said that the continuity is anywhere broken, at last leads to so much divergence of character that, but for the intermediate links, the members at each end of the chain could not be suspected of being in any way related. Well, such cases of chains of affinity obviously tell most strongly in favour of descent with continuous modification; while it is impossible to suggest why, if all the links were separately forged by as many acts of special creation, there should have been this gradual transmutation of characters carried to the point where the original creative ideal has been so completely transformed that, but for the accident of the chain being still complete, no one of nature's interpreters could possibly have discovered the connexion. For, as we have seen, this is not a case in which any appeal can be logically made to the argument from ignorance of divine method, unless some independent evidence could be adduced in favour of special creation. And that no such independent evidence exists, it will be the object of future chapters to show.

CHAPTER III.

MORPHOLOGY.

The theory of evolution supposes that hereditary characters admit of being slowly modified wherever their modification will render an organism better suited to a change in its conditions of life. Let us, then, observe the evidence which we have of such adaptive modifications of structure, in cases where the need of such modification is apparent. We may begin by again taking the case of the whales and porpoises. The theory of evolution infers, from the whole structure of these animals, that their progenitors must have been terrestrial quadrupeds of some kind, which gradually became more and more aquatic in their habits. Now the change in the conditions of their life thus brought about would have rendered desirable great modifications of structure. These changes would have begun by affecting the least typical-- that is, the least strongly inherited--structures, such as the skin, claws, and teeth. But, as time went on, the adaptation would have extended to more typical structures, until the shape of the body would have become affected by the bones and muscles required for terrestrial locomotion becoming better adapted for aquatic locomotion, and the whole outline of the animal more fish-like in shape. This is the stage which we actually observe in the seals, where the hind legs, although retaining all their typical bones, have become shortened up almost to rudiments, and directed backwards, so as to be of no use for walking, while serving to complete the fish-like taper of the body. (Fig. 2.) But in the whales the modification has gone further than this so that the hind legs have ceased to be apparent externally, and are only represented internally--and even this only in some species--by remnants so rudimentary that it is difficult to make out with certainty the homologies of the bones; moreover, the head and the whole body have become completely fish-like in shape. (Fig. 3.) But profound as are these alterations, they affect only those parts of the organism which it was for the benefit of the organism to have altered, so that it might be adapted to an aquatic mode of existence. Thus the arm, which is used as a fin, still retains the bones of the shoulder, fore-arm, wrist, and fingers, although they are all enclosed in a fin-shaped sack, so as to render them useless for any purpose other than swimming (Fig. 4.) Similarly, the head, although it so closely resembles the head of a fish in shape, still retains the bones of the mammalian skull in their proper anatomical relations to one another; but modified in form so as to offer the least possible resistance to the water. In short, it may be said that all the modifications have been effected with the least possible divergence from the typical mammalian type, which is compatible with securing so perfect an adaptation to a purely aquatic mode of life.

Now I have chosen the case of the whale and porpoise group, because they offer so extreme an example of profound modification of structure in adaptation to changed conditions of life. But the same thing may be seen in hundreds and hundreds of other cases. For instance, to confine our attention to the arm, not only is the limb modified in the whale for swimming, but in another mammal--the bat--it is modified for flying, by having the fingers enormously elongated and overspread with a membranous web.

In birds, again, the arm is modified for flight in a wholly different way--the fingers here being very short and all run together, while the chief expanse of the wing is composed of the shoulder and fore-arm. In frogs and lizards, again, we find hands more like our own; but in an extinct species of flying reptile the modification was extreme, the wing having been formed by a prodigious elongation of the fifth finger, and a membrane spread over it and the rest of the hand. (Fig. 5.) Lastly, in serpents the hand and arm have disappeared altogether.

Thus, even if we confine our attention to a single organ, how wonderful are the modifications which it is seen to undergo, although never losing its typical character. Everywhere we find the distinction between homology and analogy which was explained in the last chapter--the distinction, that is, between correspondence of structure and correspondence of function. On the one hand, we meet with structures which are perfectly homologous and yet in no way analogous: the structural elements remain, but are profoundly modified so as to perform wholly different functions. On the other hand, we meet with structures which are perfectly analogous, and yet in no way homologous: totally different structures are modified to perform the same functions. How, then, are we to explain these things? By design manifested in special creation, or by descent with adaptive modification? If it is said by design manifested in special creation, we must suppose that the Deity formed an archetypal plan of certain structures, and that he determined to adhere to this plan through all the modifications which those structures exhibit. But, if so, why is it that some structures are selected as typical and not others? Why should the vertebral skeleton, for instance, be tortured into every conceivable variety of modification in order to subserve as great a variety of functions; while another structure, such as the eye, is made in different sub-kingdoms on fundamentally different plans, notwithstanding that it has throughout to perform the same function? Will any one have the hardihood to assert that in the case of the skeleton the Deity has endeavoured to show his ingenuity, by the manifold functions to which he has made the same structure subservient; while in the case of the eye he has endeavoured to show his resources, by the manifold structures which he has adapted to serve the same function? If so, it becomes a most unfortunate circumstance that, throughout both the vegetable and animal kingdoms, all cases which can be pointed to as showing ingenious adaptation of the same typical structure to the performance of widely different functions--or cases of homology without analogy,--are cases which come within the limits of the same natural group of plants and animals, and therefore admit of being equally well explained by descent from a common ancestry; while all cases of widely different structures performing the same function--or cases of analogy without homology,--are to be found in different groups of plants or animals, and are therefore suggestive of independent variations arising in the different lines of hereditary descent.

To take a specific illustration. The octopus, or devil-fish, belongs to a widely different class of animals from a true fish; and yet its eye, in general appearance, looks wonderfully like the eye of a true fish. Now, Mr. Mivart pointed to this fact as a great difficulty in the way of the theory of evolution by natural selection, because it must clearly be a most improbable thing that so complicated a structure as the eye of a fish should happen to be arrived at through each of two totally different lines of descent. And this difficulty would, indeed, be a formidable one to the theory of evolution, if the similarity were not only analogical but homological. Unfortunately for the objection, however, Darwin clearly showed in his reply that in no one anatomical or homologous feature do the two structures resemble one another; so that, in point of fact, the two organs do not resemble one another in any particular further than it is necessary that they should, if both are to be analogous, or to serve the same function as organs of sight. But now, suppose that this had not been the case, and that the two structures, besides presenting the necessary superficial or analogical resemblance, had also presented an anatomical or homologous resemblance, with what force might it have then been urged,--Your hypothesis of hereditary descent with progressive modification being here excluded by the fact that the animals compared belong to two widely different branches of the tree of life, how are we to explain the identity of type manifested by these two complicated organs of vision? The only hypothesis open to us is intelligent adherence to an ideal plan or mechanism. But as this cannot now be urged in any comparable case throughout the whole organic world, we may on the other hand present it as a most significant fact, that while within the limits of the same large branch of the tree of life we constantly find the same typical structures modified so as to perform very different functions, we never find any of these particular types of structure in other large branches of the tree. That is to say, we never find typical structures appearing except in cases where their presence may be explained by the hypothesis of hereditary descent; while in thousands of such cases we find these structures undergoing every conceivable variety of adaptive modification.

Consequently, special creationists must fall back upon another position and say,--Well, but it may have pleased the Deity to form a certain number of ideal types, and never to have allowed the structures occurring in one type to appear in any of the others. We answer,--Undoubtedly such may have been the case; but, if so, it is a most unfortunate thing for your theory, because the fact implies that the Deity has planned his types in such a way as to suggest the counter-theory of descent. For instance, it would seem most capricious on the part of the Deity to have made the eyes of an innumerable number of fish on exactly the same ideal type, and then to have made the eye of the octopus so exactly like these other eyes in superficial appearance as to deceive so accomplished a naturalist as Mr. Mivart, and yet to have taken scrupulous care that in no one ideal particular should the one type resemble the other. However, adopting for the sake of argument this great assumption, let us suppose that God did lay down these arbitrary rules for his own guidance in creation, and then let us see to what the assumption leads. If the Deity formed a certain number of ideal types, and determined that on no account should he allow any part of one type to appear in any part of another, surely we should expect that within the limits of the same type the same typical structures should always be present. Thus, remember what efforts, so to speak, have been made to maintain the uniformity of type in the case of the fore-limb as previously explained, and should we not expect that in other and similar cases a similar method should have been followed? Yet we repeatedly find that this is not the case. Even in the whale, as we have seen, the hind-limbs are either altogether absent or dwindled almost to nothing; and it is impossible to see in what respect the hind-limbs are of any less ideal value than the fore-limbs--which are carefully preserved in all vertebrated animals except the snakes, and the extinct Dinornis, where again we meet in this particular with a sudden and sublime indifference to the maintenance of a typical structure. (Fig. 6.)[4] Now I say that if the theory of ideal types is true, we have in these facts evidence of a most unreasonable inconsistency. But the theory of descent with continued adaptive modification fully explains all the known cases; for in every case the degree of divergence from the typical structure which an organism presents corresponds, in a general way, with the length of time during which the divergence has been going on. Thus we scarcely ever meet with any great departure from the typical form with respect to one of the organs, without some of the other organs being so far modified as of themselves to indicate, on the supposition of descent with modification, that the animal or plant must have been subject to the modifying influences for an enormously long series of generations. And this combined testimony of a number of organs in the same organism is what the theory of descent would lead us to expect, while the rival theory of design can offer no explanation of the fact, that when one organ shows a conspicuous departure from the supposed ideal type, some of the other organs in the same organism should tend to keep it company by doing likewise.

[4] It is, however, probable that all species of the genus retained a tiny rudiment of wings in greatly dwindled scapulo-coracoid bones. And Mr. H. O. Forbes has detected, in a recently exhumed specimen of the latter, an indication of the glenoid cavity, for the articulation of an extremely aborted humerus. (See Nature, Jan. 14th, 1892.)

As an illustration both of this and of other points which have been mentioned, I may draw attention to what seems to me a particularly suggestive case. So-called soldier-or hermit-crabs, are crabs which have adopted the habit of appropriating the empty shells of mollusks. In association with this peculiar habit, the structure of these animals differs very greatly from that of all other crabs. In particular, the hinder part of the body, which occupies the mollusk-shell, and which therefore has ceased to require any hard covering of its own, has been suffered to lose its calcareous integument, and presents a soft fleshy character, quite unlike that of the more exposed parts of the animal. Moreover, this soft fleshy part of the creature is specially adapted to the particular requirements of the creature by having its lateral appendages--i. e. appendages which in other crustacea perform the function of legs--modified so as to act as claspers to the inside of the mollusk-shell; while the tail-end of the part in question is twisted into the form of a spiral, which fits into the spiral of the mollusk-shell. Now, in Keeling Island there is a large kind of crab called Birgus latro, which lives upon land and there feeds upon cocoa-nuts. The whole structure of this crab, it seems to me, unmistakeably resembles the structure of a hermit-crab (see drawings on the next page, Fig. 7). Yet this crab neither lives in the shell of a mollusk, nor is the hinder part of its body in the soft and fleshy condition just described: on the contrary, it is covered with a hard integument like all the other parts of the animal. Consequently, I think we may infer that the ancestors of Birgus were hermit-crabs living in mollusk-shells; but that their descendants gradually relinquished this habit as they gradually became more and more terrestrial, while, concurrently with these changes in habit, the originally soft posterior parts acquired a hard protective covering to take the place of that which was formerly supplied by the mollusk-shell. So that, if so, we now have, within the limits of a single organism, evidence of a whole series of morphological changes in the past history of its species. First, there must have been the great change from an ordinary crab to a hermit-crab in all the respects previously pointed out. Next, there must have been the change back again from a hermit-crab to an ordinary crab, so far as living without the necessity of a mollusk-shell is concerned. From an evolutionary point of view, therefore, we appear to have in the existing structure of Birgus a morphological record of all these changes, and one which gives us a reasonable explanation of why the animal presents the extraordinary appearance which it does. But, on the theory of special creation, it is inexplicable why this land-crab should have been formed on the pattern of a hermit-crab, when it never has need to enter the shell of a mollusk. In other words, its peculiar structure is not specially in keeping with its present habits, although so curiously allied to the similar structure of certain other crabs of totally different habits, in relation to which the peculiarities are of plain and obvious significance.

* * * * *

I will devote the remainder of this chapter to considering another branch of the argument from morphology, to which the case of Birgus serves as a suitable introduction: I mean the argument from rudimentary structures.

Throughout both the animal and vegetable kingdoms we constantly meet with dwarfed and useless representatives of organs, which in other and allied kinds of animals and plants are of large size and functional utility. Thus, for instance, the unborn whale has rudimentary teeth, which are never destined to cut the gums; and throughout its life this animal retains, in a similarly rudimentary condition, a number of organs which never could have been of use to any kind of creature save a terrestrial quadruped. The whole anatomy of its internal ear, for example, has reference to hearing in air--or, as Hunter long ago remarked, "is constructed upon the same principle as in the quadruped"; yet, as Owen says, "the outer opening and passage leading therefrom to the tympanum can rarely be affected by sonorous vibrations of the atmosphere, and indeed they are reduced, or have degenerated, to a degree which makes it difficult to conceive how such vibrations can be propagated to the ear-drum during the brief moments in which the opening may be raised above the water."

Now, rudimentary organs of this kind are of such frequent occurrence, that almost every species presents one or more of them--usually, indeed, a considerable number. How, then, are they to be accounted for? of course the theory of descent with adaptive modification has a simple answer to supply-- namely, that when, from changed conditions of life, an organ which was previously useful becomes useless, it will be suffered to dwindle away in successive generations, under the influence of certain natural causes which we shall have to consider in future chapters. On the other hand, the theory of special creation can only maintain that these rudiments are formed for the sake of adhering to an ideal type. Now, here again the former theory appears to be triumphant over the latter; for, without waiting to dispute the wisdom of making dwarfed and useless structures merely for the whimsical motive assigned, surely if such a method were adopted in so many cases, we should expect that in consistency it would be adopted in all cases. This reasonable expectation, however, is far from being realized. We have already seen that in numberless cases, such as that of the fore-limbs of serpents, no vestige of a rudiment is present. But the vacillating policy in the matter of rudiments does not end here; for it is shown in a still more aggravated form where within the limits of the same natural group of organisms a rudiment is sometimes present and sometimes absent. For instance, although in nearly all the numerous species of snakes there are no vestiges of limbs, in the python we find very tiny rudiments of the hind-limbs. (Fig. 8.) Now, is it a worthy conception of deity that, while neglecting to maintain his unity of ideal in the case of nearly all the numerous species of snakes, he should have added a tiny rudiment in the case of the python--and even in that case should have maintained his ideal very inefficiently, inasmuch as only two limbs, instead of four, are represented? how much more reasonable is the naturalistic interpretation; for here the very irregularity of their appearance in different species, which constitutes rudimentary structures one of the crowning difficulties to the theory of special design, furnishes the best possible evidence in favour of hereditary descent; seeing that this irregularity then becomes what may be termed the anticipated expression of progressive dwindling due to inutility. Thus, for example, to return to the case of wings, we have already seen that in an extinct genus of bird, dinornis, these organs were reduced to such an extent as to leave it still doubtful whether so much as the tiny rudiment hypothetically supplied to fig. 6 (p. 61) was present in all the species. And here is another well-known case of another genus of still existing bird, which, as was the case with dinornis, occurs only in new zealand. (Fig. 9.) Upon this island there are no four-footed enemies--either existing or extinct--to escape from which the wings of birds would be of any service. Consequently we can understand why on this island we should meet with such a remarkable dwindling away of wings.

Similarly, the logger-headed duck of South America can only flap along the surface of the water, having its wings considerably reduced though less so than the Apteryx of New Zealand. But here the interesting fact is that the young birds are able to fly perfectly well. Now, in accordance with a general law to be considered in a future chapter, the life-history of an individual organism is a kind of condensed recapitulation of the life-history of its species. Consequently, we can understand why the little chickens of the logger-headed duck are able to fly like all other ducks, while their parents are only able to flap along the surface of the water.

Facts analogous to this reduction of wings in birds which have no further use for them, are to be met with also in insects under similar circumstances. Thus, there are on the island of Madeira somewhere between 500 and 600 species of beetles, which are in large part peculiar to that island, though related to other--and therefore presumably parent--species on the neighbouring continent. Now, no less than 200 species--or nearly half the whole number--are so far deficient in wings that they cannot fly. And, if we disregard the species which are not peculiar to the island--that is to say, all the species which likewise occur on the neighbouring continent, and therefore, as evolutionists conclude, have but recently migrated to the island,--we find this very remarkable proportion. There are altogether 29 peculiar genera, and out of these no less than 23 have all their species in this condition.

Similar facts have been recently observed by the Rev. A. E. Eaton with respect to insects inhabiting Kerguelen Island. All the species which he found on the island--viz. a moth, several flies, and numerous beetles--he found to be incapable of flight; and therefore, as Wallace observes, "as these insects could hardly have reached the islands in a wingless state, even if there were any other known land inhabited by them, which there is not, we must assume that, like the Madeiran insects, they were originally winged, and lost their power of flight because its possession was injurious to them"--Kerguelen Island being "one of the stormiest places on the globe," and therefore a place where insects could rarely afford to fly without incurring the danger of being blown out to sea.

Here is another and perhaps an even more suggestive class of facts.

It is now many years ago since the editors of Silliman's Journal requested the late Professor Agassiz to give them his opinion on the following question. In a certain dark subterranean cave, called the Mammoth cave, there are found some peculiar species of blind fishes. Now the editors of Silliman's Journal wished to know whether Prof. Agassiz would hold that these fish had been specially created in these caves, and purposely devoided of eyes which could never be of any use to them; or whether he would allow that these fish had probably descended from other species, but, having got into the dark cave, gradually lost their eyes through disuse. Prof. Agassiz, who was a believer in special creation, allowed that this ought to constitute a crucial test as between the two theories of special design and hereditary descent. "If physical circumstances," he said, "ever modified organized beings, it should be easily ascertained here." And eventually he gave it as his opinion, that these fish "were created under the circumstances in which they now live, within the limits over which they now range, and with the structural peculiarities which now characterise them."

Since then a great deal of attention has been paid to the fauna of this Mammoth cave, and also to the faunas of other dark caverns, not only in the New, but also in the Old World. In the result, the following general facts have been fully established.

(1) Not only fish, but many representatives of other classes, have been found in dark caves.

(2) Wherever the caves are totally dark, all the animals are blind.

(3) If the animals live near enough to the entrance to receive some degree of light, they may have large and lustrous eyes.

(4) In all cases the species of blind animals are closely allied to species inhabiting the district where the caves occur; so that the blind species inhabiting American caves are closely allied to American species, while those inhabiting European caves are closely allied to European species.

(5) In nearly all cases structural remnants of eyes admit of being detected, in various degrees of obsolescence. In the case of some of the crustaceans of the Mammoth cave the foot-stalks of the eyes are present, although the eyes themselves are entirely absent.

Now, it is evident that all these general facts are in full agreement with the theory of evolution, while they offer serious difficulties to the theory of special creation. As Darwin remarks, it is hard to imagine conditions of life more similar than those furnished by deep limestone caverns under nearly the same climate in the two continents of America and Europe; so that, in accordance with the theory of special creation, very close similarity in the organizations of the two sets of faunas might have been expected. But, instead of this, the affinities of these two sets of faunas are with those of their respective continents--as of course they ought to be on the theory of evolution. Again, what would have been the sense of creating useless foot-stalks for the imaginary support of absent eyes, not to mention all the other various grades of degeneration in other cases? So that, upon the whole, if we agree with the late Prof. Agassiz in regarding these cave animals as furnishing a crucial test between the rival theories of creation and evolution, we must further conclude that the whole body of evidence which they now furnish is weighing on the side of evolution.

So much, then, for a few special instances of what Darwin called rudimentary structures, but what may be more descriptively designated--in accordance with the theory of descent--obsolescent or vestigial structures. It is, however, of great importance to add that these structures are of such general occurrence throughout both the vegetable and animal kingdoms, that, as Darwin has observed, it is almost impossible to point to a single species which does not present one or more of them. In other words, it is almost impossible to find a single species which does not in this way bear some record of its own descent from other species; and the more closely the structure of any species is examined anatomically, the more numerous are such records found to be. Thus, for example, of all organisms that of man has been most minutely investigated by anatomists; and therefore I think it will be instructive to conclude this chapter by giving a list of the more noteworthy vestigial structures which are known to occur in the human body. I will take only those which are found in adult man, reserving for the next chapter those which occur in a transitory manner during earlier periods of his life. But, even as thus restricted, the number of obsolescent structures which we all present in our own persons is so remarkable, that their combined testimony to our descent from a quadrumanous ancestry appears to me in itself conclusive. I mean, that even if these structures stood alone, or apart from any more general evidences of our family relationships, they would be sufficient to prove our parentage. Nevertheless, it is desirable to remark that of course these special evidences which I am about to detail do not stand alone. Not only is there the general analogy furnished by the general proof of evolution elsewhere, but there is likewise the more special correspondence between the whole of our anatomy and that of our nearest zoological allies. Now the force of this latter consideration is so enormous, that no one who has not studied human anatomy can be in a position to appreciate it. For without special study it is impossible to form any adequate idea of the intricacy of structure which is presented by the human form. Yet it is found that this enormously intricate organization is repeated in all its details in the bodies of the higher apes. There is no bone, muscle, nerve, or vessel of any importance in the one which is not answered to by the other. Hence there are hundreds of thousands of instances of the most detailed correspondence, without there being any instances to the contrary, if we pay due regard to vestigial characters. The entire corporeal structure of man is an exact anatomical copy of that which we find in the ape.

My object, then, here is to limit attention to those features of our corporeal structure which, having become useless on account of our change in attitude and habits, are in process of becoming obsolete, and therefore occur as mere vestigial records of a former state of things. For example, throughout the vertebrated series, from fish to mammals, there occurs in the inner corner of the eye a semi-transparent eye-lid, which is called the nictitating membrane. The object of this structure is to sweep rapidly, every now and then, over the external surface of the eye, apparently in order to keep the surface clean. But although the membrane occurs in all classes of the sub-kingdom, it is more prevalent in some than in others--e.g. in birds than in mammals. Even, however, where it does not occur of a size and mobility to be of any use, it is usually represented, in animals above fishes, by a functionless rudiment, as here depicted in the case of man. (Fig. 10.)

Now the organization of man presents so many vestigial structures thus referring to various stages of his long ancestral history, that it would be tedious so much as to enumerate them. Therefore I will yet further limit the list of vestigial structures to be given as examples, by not only restricting these to cases which occur in our own organization; but of them I shall mention only such as refer us to the very last stage of our ancestral history--viz. structures which have become obsolescent since the time when our distinctively human branch of the family tree diverged from that of our immediate forefathers, the Quadrumana.

(1) Muscles of the external ear.--These, which are of large size and functional use in quadrupeds, we retain in a dwindled and useless condition (Fig. 11). This is likewise the case in anthropoid apes; but in not a few other Quadrumana (e.g. baboons, macacus, magots, &c.) degeneration has not proceeded so far, and the ears are voluntarily moveable.

(2) Panniculus carnosis.--A large number of the mammalia are able to move their skin by means of sub-cutaneous muscle--as we see, for instance, in a

horse, when thus protecting himself against the sucking of flies. We, in common with the Quadrumana, possess an active remnant of such a muscle in the skin of the forehead, whereby we draw up the eyebrows; but we are no longer able to use other considerable remnants of it, in the scalp and elsewhere,--or, more correctly, it is rarely that we meet with persons who can. But most of the Quadrumana (including the anthropoids) are still able to do so. There are also many other vestigial muscles, which occur only in a small percentage of human beings, but which, when they do occur, present unmistakeable homologies with normal muscles in some of the Quadrumana and still lower animals[5].

[5] See especially Mr. John Wood's papers, Proc. R. S., xiii to xvi, and xviii; also Journ. Anat., i and iii. In this connexion Darwin refers to M. Richard, Annls. d. Sc. Nat. Zoolg., tom. xviii, p. 13, 1852.

(3) Feet.--It is observable that in the infant the feet have a strong deflection inwards, so that the soles in considerable measure face one another. This peculiarity, which is even more marked in the embryo than in the infant (see p. 153), and which becomes gradually less and less conspicuous even before the child begins to walk, appears to me a highly suggestive peculiarity. For it plainly refers to the condition of things in the Quadrumana, seeing that in all these animals the feet are similarly curved inwards, to facilitate the grasping of branches. And even when walking on the ground apes and monkeys employ to a great extent the outside edges of their feet, as does also a child when learning to walk. The feet of a young child are also extraordinarily mobile in all directions, as are those of apes. In order to show these points, I here introduce comparative drawings of a young ape and the portrait of a young male child. These drawings, moreover, serve at the same time to illustrate two other vestigial characters, which have often been previously noticed with regard to the infant's foot. I allude to the incurved form of the legs, and the lateral extension of the great toe, whereby it approaches the thumb-like character of this organ in the Quadrumana. As in the case of the incurved position of the legs and feet, so in this case of the lateral extensibility of the great toe, the peculiarity is even more marked in embryonic than in infant life. For, as Prof. Wyman has remarked with regard to the foetus when about an inch in length, "The great toe is shorter than the others; and, instead of being parallel to them, is projected at an angle from the side of the foot, thus corresponding with the permanent condition of this

part in the Quadrumana[6]." So that this organ, which, according to Owen, "is perhaps the most characteristic peculiarity in the human structure," when traced back to the early stages of its development, is found to present a notably less degree of peculiarity.

[6] Proc. Nat. Hist. Soc., Boston, 1863.

(4) Hands.--Dr. Louis Robinson has recently observed that the grasping power of the whole human hand is so surprisingly great at birth, and during the first few weeks of infancy, as to be far in excess of present requirements on the part of a young child. Hence he concludes that it refers us to our quadrumanous ancestry--the young of anthropoid apes being endowed with similar powers of grasping, in order to hold on to the hair of the mother when she is using her arms for the purposes of locomotion. This inference appears to me justifiable, inasmuch as no other explanation can be given of the comparatively inordinate muscular force of an infant's grip. For experiments showed that very young babies are able to support their own weight, by holding on to a horizontal bar, for a period varying from one half to more than two minutes[7]. With his kind permission I here reproduce one of Dr. Robinson's instantaneous, and hitherto unpublished, photographs of a very young infant. This photograph was taken after the above paragraph (3) was written, and I introduce it here because it serves to show incidentally--and perhaps even better than the preceding figure--the points there mentioned with regard to the feet and great toes. Again, as Dr. Robinson observes, the attitude, and the disproportionately large development of the arms as compared with the legs, give all the photographs a striking resemblance to a picture of the chimpanzee "Sally" at the Zoological Gardens. For "invariably the thighs are bent nearly at right angles to the body, and in no case did the lower limbs hang down and take the attitude of the erect position." He adds, "In many cases no sign of distress is evinced, and no cry uttered, until the grasp begins to give way."

[7] Nineteenth Century, November, 1891.

(5) Tail.--The absence of a tail in man is popularly supposed to constitute a difficulty against the doctrine of his quadrumanous descent. As a matter of fact, however, the absence of an external tail in man is precisely what this doctrine would expect, seeing that the nearest allies of man in the

quadrumanous series are likewise destitute of an external tail. Far, then, from this deficiency in man constituting any difficulty to be accounted for, if the case were not so--i. e. if man did possess an external tail,--the difficulty would be to understand how he had managed to retain an organ which had been renounced by his most recent ancestors. Nevertheless, as the anthropoid apes continue to present the rudimentary vestiges of a tail in a few caudal vertebr?below the integuments, we might well expect to find a similar state of matters in the case of man. And this is just what we do find, as a glance at these two comparative illustrations will show. (Fig. 15.) Moreover, during embryonic life, both of the anthropoid apes and of man, the tail much more closely resembles that of the lower kinds of quadrumanous animals from which these higher representatives of the group have descended. For at a certain stage of embryonic life the tail, both of apes and of human beings, is actually longer than the legs (see Fig. 16). And at this stage of development, also, the tail admits of being moved by muscles which later on dwindle away. Occasionally, however, these muscles persist, and are then described by anatomists as abnormalities. The following illustrations serve to show the muscles in question, when thus found in adult man.

(6) Vermiform Appendix of the Cecum.--This is of large size and functional use in the process of digestion among many herbivorous animals; while in man it is not only too small to serve any such purpose, but is even a source of danger to life--many persons dying every year from inflammation set up by the lodgement in this blind tube of fruit-stones, &c.

In the orang it is longer than in man (Fig. 18), as it is also in the human foetus proportionally compared with the adult. (Fig. 19.) In some of the lower herbivorous animals it is longer than the entire body.

Like vestigial structures in general, however, this one is highly variable. Thus the above cut (Fig. 19) serves to show that it may sometimes be almost as short in the orang as it normally is in man--both the human subjects of this illustration having been normal.

(7) Ear.--Mr. Darwin writes:--

The celebrated sculptor, Mr. Woolner, informs me of one little peculiarity in the external ear, which he has often observed both in men and women.... The peculiarity consists in a little blunt point, projecting from the inwardly folded margin, or helix. When present, it is developed at birth, and, according to Prof. Ludwig Meyer, more frequently in man than in woman. Mr. Woolner made an exact model of one such case, and sent me the accompanying drawing.... The helix obviously consists of the extreme margin of the ear folded inwards; and the folding appears to be in some manner connected with the whole external ear being permanently pressed backwards. In many monkeys, which do not stand high in the order, as baboons and some species of macacus, the upper portion of the ear is slightly pointed, and the margin is not at all folded inwards; but if the margin were to be thus folded, a slight point would necessarily project towards the centre.... The following wood-cut is an accurate copy of a photograph of the foetus of an orang (kindly sent me by Dr. Nitsche), in which it may be seen how different the pointed outline of the ear is at this period from its adult condition, when it bears a close general resemblance to that of man [including even the occasional appearance of the projecting point shown in the preceding woodcut]. It is evident that the folding over of the tip of such an ear, unless it changed greatly during its further development, would give rise to a point projecting inwards[8].

[8] Descent of Man, 2nd ed., pp. 15-16.

The following woodcut serves still further to show vestigial resemblances between the human ear and that of apes. The last two figures illustrate the general resemblance between the normal ear of foetal man and the ear of an adult orang-outang. The other two figures on the lower line are intended to exhibit occasional modifications of the adult human ear, which approximate simian characters somewhat more closely than does the normal type. It will be observed that in their comparatively small lobes these ears resemble those of all the apes; and that while the outer margin of one is not unlike that of the Barbary ape, the outer margin of the other follows those of the chimpanzee and orang. Of course it would be easy to select individual human ears which present either of these characters in a more pronounced degree; but these ears have been chosen as models because they present both characters in conjunction. The upper row of figures likewise shows the close similarity of hair-tracts, and the direction of growth on the part of the hair itself, in cases where the human ear happens to be of an abnormally hirsute character. But this particular instance (which I do not think has been previously noticed) introduces us to the subject of hair, and hair-growth, in general.

(8) Hair.--Adult man presents rudimentary hair over most parts of the body. Wallace has sought to draw a refined distinction between this vestigial coating and the useful coating of quadrumanous animals, in the absence of the former from the human back. But even this refined distinction does not hold. On the one hand, the comparatively hairless chimpanzee which died last year in the Zoological Gardens (T. calvus) was remarkably denuded over the back; and, on the other hand, men who present a considerable development of hair over the rest of their bodies present it also on their backs and shoulders. Again, in all men the rudimentary hair on the upper and lower arm is directed towards the elbow--a peculiarity which occurs nowhere else in the animal kingdom, with the exception of the anthropoid apes and a few American monkeys, where it presumably has to do with arboreal habits. For, when sitting in trees, the orang, as observed by Mr. Wallace, places its hands above its head with its elbows pointing downwards: the disposition of hair on the arms and fore-arms then has the effect of thatch in turning the rain. Again, I find that in all species of apes, monkeys, and baboons which I have examined (and they have been numerous), the hair on the backs of the hands and feet is continued as far as the first row of phalanges; but becomes scanty, or disappears altogether, on the second row; while it is invariably absent on the terminal row. I also find that the same peculiarity occurs in man. We all have rudimentary hair on the first row of phalanges, both of hands and feet: when present at all, it is more scanty on the second row; and in no case have I been able to find any on the terminal row. In all cases these peculiarities are congenital, and the total absence or partial presence of hair on the second phalanges is constant in different species of Quadrumana. For instance, it is entirely absent in all the chimpanzees, which I have examined, while scantily present in all the orangs. As in man, it occurs in a patch midway between the joints.

Besides showing these two features with regard to the disposition of hair on the human arm and hand, the above woodcut illustrates a third. By looking closely at the arm of the very hairy man from whom the drawing was taken, it could be seen that there was a strong tendency towards a whorled arrangement of the hairs on the backs of the wrists. This is likewise, as a general rule, a marked feature in the arrangement of hair on the same places in the gorilla, orang, and chimpanzee. In the specimen of the latter, however, from which the drawing was taken, this characteristic was not well marked. The downward direction of the hair on the backs of the hands is exactly the same in man as it is in all the anthropoid apes. Again, with regard to hair, Darwin notices that occasionally there appears in man a few hairs in the eyebrows much longer than the others; and that they seem to be representative of similarly long and scattered hairs which occur in the chimpanzee, macacus, and baboons.

Lastly, it may be here more conveniently observed than in the next chapter on Embryology, that at about the sixth month the human foetus is often thickly coated with somewhat long dark hair over the entire body, except the soles of the feet and palms of the hands, which are likewise bare in all quadrumanous animals. This covering, which is called the lanugo, and sometimes extends even to the whole forehead, ears, and face, is shed before birth. So that it appears to be useless for any purpose other than that of emphatically declaring man a child of the monkey.

(9) Teeth.--Darwin writes:--

It appears as if the posterior molar or wisdom-teeth were tending to become rudimentary in the more civilized races of man. These teeth are rather smaller than the other molars, as is likewise the case with the corresponding teeth in the chimpanzee and orang; and they have only two separate fangs.... They are also much more liable to vary, both in structure and in the period of their development, than the other teeth. In the Melanian races, on the other hand, the wisdom-teeth are usually furnished with three separate fangs, and are usually sound [i. e. not specially liable to decay]; they also differ from the other molars in size, less than in the Caucasian races.

Now, in addition to these there are other respects in which the dwindling condition of wisdom-teeth is manifested--particularly with regard to the pattern of their crowns. Indeed, in this respect it would seem that even in the anthropoid apes there is the beginning of a tendency to degeneration of the molar teeth from behind forwards. For if we compare the three molars in the lower jaw of the gorilla, orang, and chimpanzee, we find that the gorilla has five well-marked cusps on all three of them; but that in the orang the cusps are not so pronounced, while in the chimpanzee there are only four of them on the third molar. Now in man it is only the first of these three teeth which normally presents five cusps, both the others presenting only four. So that, comparing all these genera together, it appears that the number of cusps is being reduced from behind forwards; the chimpanzee having lost one of them from the third molar, while man has not only lost this, but also one from the second molar,--and, it may be added, likewise partially (or even totally) from the first molar, as a frequent variation among civilized races. But, on the other hand, variations are often met with in the opposite direction, where the second or the third molar of man presents five cusps--in the one case following the chimpanzee, in the other the gorilla. These latter variations, therefore, may fairly be regarded as reversionary. For these facts I am indebted to the kindness of Mr. C. S. Tomes.

(10) Perforations of the humerus.--The peculiarities which we have to notice under this heading are two in number. First, the supra condyloid foramen is a normal feature in some of the lower Quadrumana (Fig. 25), where it gives passage to the great nerve of the fore-arm, and often also to the great artery. In man, however, it is not a normal feature. Yet it occurs in a small percentage of cases--viz., according to Sir W. Turner, in about one per cent., and therefore is regarded by Darwin as a vestigial character. Secondly, there is inter-condyloid foramen, which is also situated near the lower end of the humerus, but more in the middle of the bone. This occurs, but not constantly, in apes, and also in the human species. From the fact that it does so much more frequently in the bones of ancient--and also of some savage--races of mankind (viz. in 20 to 30 per cent. of cases), Darwin is disposed to regard it also as a vestigial feature. On the other hand, Prof. Flower tells me that in his opinion it is but an expression of impoverished nutrition during the growth of the bone.

(11) Flattening of tibia.--In some very ancient human skeletons, there has also been found a lateral flattening of the tibia, which rarely occurs in any existing human beings, but which appears to have been usual among the earliest races of mankind hitherto discovered. According to Broca, the measurements of these fossil human tibi?resemble those of apes. Moreover, the bone is bent and strongly convex forwards, while its angles are so rounded as to present the nearly oval section seen in apes. It is in association with these ape-like human tibi?that perforated humeri of man are found in greatest abundance.

On the other hand, however, there is reason to doubt whether this form of tibia in man is really a survival from his quadrumanous ancestry. For, as Boyd-Dawkins and Hartmann have pointed out, the degree of flattening presented by some of these ancient human bones is greater than that which occurs in any existing species of anthropoid ape. Of course the possibility remains that the unknown species of ape from which man descended may have had its tibia more flattened than is now observable in any of the existing species. Nevertheless, as some doubt attaches to this particular case, I do not press it--and, indeed, only mention it at all in order that the doubt may be expressed.

Similarly, I will conclude by remarking that several other instances of the survival of vestigial structures in man have been alleged, which are of a still more doubtful character. Of such, for example, are the supposed absence of the genial tubercle in the case of a very ancient jaw-bone of man, and the disposition of valves in human veins. From the former it was argued that the possessor of this very ancient jaw-bone was probably speechless, inasmuch as the tubercle in existing man gives attachment to muscles of the tongue. From the latter it has been argued that all the valves in the veins of the human body have reference, in their disposition, to the incidence of blood-pressure when the attitude of the body is horizontal, or quadrupedal. Now, the former case has already broken down, and I find that the latter does not hold. But we can well afford to lose such doubtful and spurious cases, in view of all the foregoing unquestionable and genuine cases of vestigial structures which are to be met with even within the limits of our own organization--and even when these limits are still further limited by selecting only those instances which refer to the very latest chapter of our long ancestral history.

CHAPTER IV.

EMBRYOLOGY.

We will next consider what of late years has become the most important of the lines of evidence, not only in favour of the general fact of evolution, but also of its history: I mean the evidence which has been yielded by the newest of the sciences, the science of Embryology. But here, as in the analogous case of adult morphology, in order to do justice to the mass of evidence which has now been accumulated, a whole volume would be necessary. As in that previous case, therefore, I must restrict myself to giving an outline sketch of the main facts.

First I will display what in the language of Paley we may call "the state of the argument."

It is an observable fact that there is often a close correspondence between developmental changes as revealed by any chronological series of fossils which may happen to have been preserved, and developmental changes which may be observed during the life-history of now existing individuals belonging to the same group of animals. For instance, the successive development of prongs in the horns of deer-like animals, which is so clearly shown in the geological history of this tribe, is closely reproduced in the life-history of existing deer. Or, in other words, the antlers of an existing deer furnish in their development a kind of summary or recapitulation, of the successive phases whereby the primitive horn was gradually superseded by horns presenting a greater and greater number of prongs in successive species of extinct deer (Fig. 26). Now it must be obvious that such a recapitulation in the life-history of an existing animal of developmental changes successively distinctive of sundry allied, though now extinct species, speaks strongly in favour of evolution. For as it is of the essence of this theory that new forms arise from older forms by way of hereditary descent, we should antecedently expect, if the theory is true, that the phases of development presented by the individual organism would follow, in their main outlines, those phases of development through which their long line of ancestors had passed. The only alternative view is that as species of deer, for instance, were separately created, additional prongs were successively added to their antlers; and yet that, in order to be so added to successive species every individual deer belonging to later species was required to repeat in his own lifetime the process of successive additions which had previously taken place in a remote series of extinct species. Now I do not deny that this view is a possible view; but I do deny that it is a probable one. According to the evolutionary interpretation of such facts, we can see a very good reason why the life-history of the individual is thus a condensed summaryof the life-history of its ancestral species. But according to the opposite view no reason can be assigned why such should be the case. In a previous chapter--the chapter on Classification--we have seen that if each species were created separately, no reason can be assigned why they should all have been turned out upon structural patterns so strongly suggestive of hereditary descent with gradual modifications, or slow divergence--the result being group subordinated to group, with the most generalized (or least developed) forms at the bottom, and the highest products of organization at the top. And now we see--or shall immediately see--that this consideration admits of being greatly fortified by a study of the developmental history of every individual organism. If it would be an unaccountable fact that every separately created species should have been created with close structural resemblances to a certain limited number of other species, less close resemblances to certain further species, and so backwards; assuredly it would be a still more unaccountable fact that every individual of every species should exhibit in its own person a history of developmental change, every term of which corresponds with the structural peculiarities of its now extinct predecessors-- and this in the exact historical order of their succession in geological time. The more that we think about this antithesis between the naturalistic and the non-naturalistic interpretations, the greater must we feel the contrast in respect of rationality to become; and, therefore, I need not spend time by saying anything further upon the antecedent standing of the two theories in this respect. The evidence, then, which I am about to adduce from the study of development in the life-histories of individual organisms, will be regarded by me as so much unquestionable evidence in favour of similar processes of development in the life-histories of their respective species--in so far, I mean, as the two sets of changes admit of being proved parallel.

[Illustration: FIG. 26.--Antlers of Stag, showing successive addition of branches in successive years. Drawn from nature (Brit. Mus.).]

In the only illustration hitherto adduced--viz. that of deers' horns--the series of changes from a one-pronged horn to a fully developed arborescent antler, is a series which takes place during the adult life of the animal; for it is only when the breeding age has been attained that horns are required to appear. But seeing that every animal passes through most of the phases of its development, not only before the breeding age has been attained, but even before the time of its own birth, clearly the largest field for the study of individual development is furnished by embryology. For instance, there is a salamander which differs from most other salamanders in being exclusively terrestrial in its habits. Now, the young of this salamander before their birth are found to be furnished with gills, which, however, they are never destined to use. Yet these gills are so perfectly formed, that if the young salamanders be removed from the body of their mother shortly before birth, and be then immediately placed in water, the little animals show themselves quite capable of aquatic respiration, and will merrily swim about in a medium which would quickly drown their own parent. Here, then, we have both morphological and physiological evidence pointing to the possession of gills by the ancestors of the land salamander.

It would be easy to devote the whole of the present chapter to an enumeration of special instances of the kinds thus chosen for purposes of illustration; but as it is desirable to take a deeper, and therefore a more general view of the whole subject, I will begin at the foundation, and gradually work up from the earliest stages of development to the latest. Before starting, however, I ask the reader to bear in mind one consideration, which must reasonably prevent our anticipating that in every case the life-history of an individual organism should present a full recapitulation of the life-history of its ancestral line of species. Supposing the theory of evolution to be true, it must follow that in many cases it would have been more or less disadvantageous to a developing type that it should have been obliged to reproduce in its individual representatives all the phases of development previously undergone by its ancestry--even within the limits of the same family. We can easily understand, for example, that the waste of material required for building up the useless gills of the embryonic salamanders is a waste which, sooner or later, is likely to be done away with; so that the fact of its occurring at all is in itself enough to show that the change from aquatic to terrestrial habits on the part of this species must have been one of comparatively recent occurrence. Now, in as far as it is detrimental to a developing type that it should pass through any particular ancestral phases of development, we may be sure that natural selection--or whatever other adjustive causes we may suppose to have been at work in the adaptation of organisms to their surroundings--will constantly seek to get rid of this necessity, with the result, when successful, of dropping out the detrimental phases. Thus the foreshortening of developmental history which takes place in the individual lifetime may be expected often to take place, not only in the way of condensation, but also in the way of excision. Many pages of ancestral history may be recapitulated in the paragraphs of embryonic development, while others may not be so much as mentioned. And that this is the true explanation of what embryologists term "direct" development--or of a more or less sudden leap from one phase to another, without any appearance of intermediate phases--is proved by the fact that in some cases both direct and indirect development occur within the same group of organisms, some genera or families having dropped out the intermediate phases which other genera or families retain.

* * * * *

The argument from embryology must be taken to begin with the first beginning of individual life in the ovum. And, in order to understand the bearings of the argument in this its first stage, we must consider the phenomena of reproduction in the simplest form which these phenomena are known to present.

The whole of the animal kingdom is divided into two great groups, which are called the Protozoa and the Metazoa. Similarly, the whole of the vegetable kingdom is divided into the Protophyta and the Metaphyta. The characteristic feature of all the Protozoa and Protophyta is that the organism consists of a single physiological cell, while the characteristic of all the Metazoa and Metaphyta is that the organism consists of a plurality of physiological cells, variously modified to subserve different functions in the economy of the animal or plant, as the case may be. For the sake of brevity, I shall hereafter deal only with the case of animals (Protozoa and Metazoa); but it may throughout be understood that everything which is said applies also to the case of plants (Protophyta and Metaphyta).

A Protozoa (like a Protophyton) is a solitary cell, or a "unicellular organism," while a Metazoa (like a Metaphyton) is a society of cells, or a "multicellular organism." Now, it is only in the multicellular organisms that there is any observable distinction of sex. In all the unicellular organisms the phenomena of reproduction appear to be more or less identical with those of growth. Nevertheless, as these phenomena are here in some cases suggestively peculiar, I will consider them more in detail.

A Protozoa is a single corpuscle of protoplasm which in different species of Protozoa varies in size from more than one inch to less than 1/1000 of an inch in diameter. In some species there is an enveloping cortical substance; in other species no such substance can be detected. Again, in most species there is a nucleus, while in other species no such differentiation of structure has hitherto been observed. Nevertheless, from the fact that the nucleus occurs in the majority of Protozoa, coupled with the fact that the demonstration of this body is often a matter of extreme difficulty, not only in some of the Protozoa where it has been but recently detected, but also in the case of certain physiological cells elsewhere,--from these facts it is not unreasonable to suppose that all the Protozoa possess a nucleus, whether or not it admits of being rendered visible by histological methods thus far at our disposal. If this is the case, we should be justified in saying, as I have said, that a Protozoa is an isolated physiological cell, and, like cells in general, multiplies by means of what Spencer and Häckel have aptly called a process of discontinuous growth. That is to say, when a cell reaches maturity, further growth takes place in the direction of a severance of its substance--the separated portion thus starting anew as a distinct physiological unit. But, notwithstanding the complex changes which have been more recently observed to take place in the nucleus of some Protozoa prior to their division, the process of multiplication by division may still be regarded as a process of growth, which differs from the previous growth of the individual cell in being attended by a severance of continuity. If we take a suspended drop of gum, and gradually add to its size by allowing more and more gum to flow into it, a point will eventually be reached at which the force of gravity will overcome that of cohesion, and a portion of the drop will fall away from the remainder. Here we have a rough physical simile, although of course no true analogy. In virtue of a continuous assimilation of nutriment, the protoplasm of a cell increases in mass, until it reaches the size at which the forces of disruption overcome those of cohesion--or, in other words, the point at which increase of size is no longer compatible with continuity of substance. Nevertheless, it must not be supposed that the process is thus merely a physical one. The phenomena which occur even in the simplest--or so-called "direct"--cell-division, are of themselves enough to prove that the process is vital, or physiological; and this in a high degree of specialization. But so, likewise, are all processes of growth in organic structures; and therefore the simile of the drop of gum is not to be regarded as a true analogy: it serves only to indicate the fact that when cell-growth proceeds beyond a certain point cell-division ensues. The size to which cells may grow before they thus divide is very variable in different kinds of cells; for while some may normally attain a length of ten or twelve inches, others divide before they measure 1/1000 of an inch. This, however, is a matter of detail, and does not affect the general physiological principles on which we are at present engaged.

Now, as we have seen, a Protozoa is a single cell; for even although in some of the higher forms of protozoal life a colony of cells may be bound together in organic connexion, each of these cells is in itself an "individual," capable of self-nourishment, reproduction, and, generally, of independent existence. Consequently, when the growth of a Protozoa ends in a division of its substance, the two parts wander away from each other as separate organisms. (Fig. 27.)

The next point we have to observe is, that in all cases where a cell or a Protozoa multiplies by way of fissiparous division, the process begins in the nucleus. If the nucleus divides into two parts, the whole cell will eventually divide into two parts, each of which retains a portion of the original nucleus, as represented in the above figure. If the nucleus divides into three, four, or even, as happens in the development of some embryonic tissues, into as many as six parts, the cell will subdivide into a corresponding number, each retaining a portion of the nucleus. Therefore, in all cases of fissiparous division, the seat or origin of the process is the nucleus.

Thus far, then, the phenomena of multiplication are identical in all the lowest or unicellular organisms, and in the constituent cells of all the higher or multicellular. And this is the first point which I desire to make apparent. For where the object is to prove a continuity between the phenomena of growth and reproduction, it is of primary importance to show--1st, that there is such a continuity in the case of all the unicellular organisms, and, 2nd, that there are all the above points of resemblance between the multiplication of cells in the unicellular and in the multicellular organisms.

It remains to consider the points of difference, and, if possible, to show that these do not go to disprove the doctrine of continuity which the points of resemblance so forcibly indicate.

The first point of difference obviously is, that in the case of all the multicellular organisms the two or more "daughter-cells," which are produced by division of the "mother-cell," do not wander away from one another; but, as a rule, they continue to be held in more or less close apposition by means of other cells and binding membranes,--with the result of giving rise to those various "tissues," which in turn go to constitute the material of "organs." I cannot suppose, however, that any advocate of discontinuity will care to take his stand at this point. But, if any one were so foolish as to do so, it would be easy to dislodge him by describing the state of matters in some of the Protozoa where a number of unicellular "individuals" are organically united so as to form a "colony." These cases serve to bridge this distinction between Protozoa and Metazoa, of which therefore we may now take leave.

In the second place, there is the no less obvious distinction that the result of cell-division in the Metazoa is not merely to multiply cells all of the same kind: on the contrary, the process here gives rise to as many different kinds of cells as there are different kinds of tissue composing the adult organism. But no one, I should think, is likely to oppose the doctrine of continuity on the ground of this distinction. For the distinction is clearly one which must necessarily arise, if the doctrine of continuity between unicellular and multicellular organisms be true. In other words, it is a distinction which the theory of evolution itself must necessarily pre-suppose, and therefore it is no objection to the theory that its pre-supposition is realized. Moreover, as we shall see better presently, there is no difficulty in understanding why this distinction should have arisen, so soon as it became necessary (or desirable) that individual cells, when composing a "colony," should conform to the economic principle of the division of labour--a principle, indeed, which is already foreshadowed in the constituent parts of a single cell, since the nucleus has one set of functions and its surrounding protoplasm another.

But now, in the third place, we arrive at a more important distinction, and one which lies at the root of the others still remaining to be considered. I refer to sexual propagation. For it is a peculiarity of the multicellular organisms that, although many of them may likewise propagate themselves by other means (Fig. 28), they all propagate themselves by means of sexual congress. Now, in its essence, sexual congress consists in the fusion of two specialized cells (or, as now seems almost certain, of the nuclei thereof), so that it is out of such a combination that the new individual arises by means of successive cell-divisions, which, beginning in the fertilized ovum, eventually build up all the tissues and organs of the body.

This process clearly indicates very high specialization on the part of germ-cells. For we see by it that although these cells when young resemble all other cells in being capable of self-multiplication by binary division (thus reproducing cells exactly like themselves), when older they lose this power; but, at the same time, they acquire an entirely new and very remarkable power of giving rise to a vast succession of many different kinds of cells, all of which are mutually correlated as to their several functions, so as to constitute a hierarchy of cells--or, to speak literally, a multicellular co-organization. Here it is that we touch the really important distinction between the Protozoa and the Metazoa; for although I have said that some of the higher Protozoa foreshadow this state of matters in forming cell-colonies, it must now be noted that the cells composing such colonies are all of the same kind; and, therefore, that the principle of producing different kinds of cells which, by mutual co-adaptation of functions, shall be capable of constructing a multicellular Metazoa,--this great principle of co-organization is but dimly nascent in the cell-colonies of Protozoa. And its marvellous development in the Metazoa appears ultimately to depend upon the highly specialized character of germ-cells. Even in cases where multicellular organisms are capable of reproducing their kind without the need of any preceding process of fertilization (parthenogenesis), and even in the still more numerous cases where complete organisms are budded forth from any part of their parent organism (gemmation, Fig. 28), there is now very good reason to conclude that these powers of a-sexual reproduction on the part of multicellular organisms are all ultimately due to the specialized character of their germ-cells. For in all these cases the tissues of the parent, from which the budding takes place, were ultimately derived from germ-cells--no matter how many generations of budded organisms may have intervened. And that propagation by budding, &c, in multicellular organisms is thus ultimately due to their propagation by sexual methods, seems to be further shown by certain facts which will have to be discussed at some length in my next volume. Here, therefore, I will mention only one of them--and this because it furnishes what appears to be another important distinction between the Protozoa and the Metazoa.

In nearly all cases where a Protozoa multiplies itself by fission, the process begins by a simple division of the nucleus. But when a Metazoa is developed from a germ-cell, although the process likewise begins by a division of the nucleus, this division is not a simple or direct one; on the contrary, it is inaugurated by a series of processes going on within the nucleus, which are so enormously complex, and withal so beautifully ordered, that to my mind they constitute the most wonderful--if not also the most suggestive--which have ever been revealed by microscopical research. It is needless to say that I refer to the phenomena of karyokinesis. A few pages further on they will be described more fully. For our present purposes it is sufficient to give merely a pictorial illustration of their successive phases; for a glance at such a representation serves to reveal the only point to which attention has now to be drawn--namely, the immense complexity of the processes in question, and therefore the contrast which they furnish to the simple (or "direct") division of the nucleus preparatory to cell-division in the unicellular organisms. Here, then (Fig. 29), we see the complex processes of karyokinesis in the first two stages of egg-cell division. But similar processes continue to repeat themselves in subsequent stages; and this, there is now good reason to believe, throughout all the stages of cell-division, whereby the original egg-cell eventually constructs an entire organism. In other words, all the cells composing all the tissues of a multicellular organism, at all stages of its development, are probably originated by these complex processes, which differ so much from the simple process of direct division in the unicellular organisms[9]. In this important respect, therefore, it does at first sight appear that we have a distinction between the Protozoa and the Metazoa of so pronounced a character, as fairly to raise the question whether cell-division is fundamentally identical in unicellular and in multicellular organisms.

[9] I say "probably," because analogy points in this direction. As a matter of fact, in many cases of tissue-formation karyokinesis has not hitherto been detected. But even if in such cases it does not occur--i. e. if failure to detect its occurrence be not due merely to still remaining imperfections of our histological methods,--the large number of cases in which it has been seen to occur in the formation of sundry tissues are of themselves sufficient to indicate some important difference between cells derived from ova (metazoal), and cells which have not been so derived (protozoal). Which is the point now under discussion.

Lastly, the only other distinction of a physiologically significant kind between a single cell when it occurs as a Protozoa and when it does so as the unfertilized ovum of a Metazoa is, that in the latter case the nucleus discharges from its own substance two minute protoplasmic masses ("polar bodies"), which are then eliminated from the cell altogether. This process, which will be more fully described later on, appears to be of invariable occurrence in the case of all egg-cells, while nothing resembling it has ever been observed in any of the Protozoa.

We must now consider these several points of difference seriatim.

First, with regard to sexual propagation, we have already seen that this is by no means the only method of propagation among the multicellular organisms; and it now remains to add that, on the other hand, there is, to say the least, a suggestive foreshadowing of sexual propagation among the unicellular organisms. For although simple binary fission is here the more usual mode of multiplication, very frequently two (rarely three or more) Protozoa of the same species come together, fuse into a single mass, and thus become very literally "one flesh." This process of "conjugation" is usually (though by no means invariably) followed by a period of quiescent "encystation"; after which the contents of the cyst escape in the form of a number of minute particles, or "spores," and these severally develope into the parent type. Obviously this process of conjugation, when it is thus a preliminary to multiplication, appears to be in its essence the same as fertilization. And if it be objected that encystation and spore-formation in the Protozoa are not always preceded by conjugation, the answer would be that neither is oviparous propagation in the Metazoa invariably preceded by fertilization.

Nevertheless, that there are great distinctions between true sexual propagation and this foreshadowing of it in conjugation I do not deny. The question, however, is whether they be so great as to justify any argument against an historical continuity between them. What, then, are these remaining distinctions? Briefly, as we have seen, they are the extrusion from egg-cells of polar bodies, and the occurrence, both in egg-cells and their products (tissue-cells), of the process of karyokinesis. But, as regards the polar bodies, it is surely not difficult to suppose that, whatever their significance may be, it is probably in some way or another connected with the high specialization of the functions which an egg-cell has to discharge. Nor is there any difficulty in further supposing that, whatever purpose is served by getting rid of polar bodies, the process whereby they are got rid of was originally one of utilitarian development--i. e. a process which at its commencement did not betoken any difference of kind, or breach of continuity, between egg-cells and cells of simpler constitution.

Lastly, with respect to karyokinesis, although it is true that the microscope has in comparatively recent years displayed this apparently important distinction between unicellular and multicellular organisms, two considerations have here to be supplied. The first is, that in some of the Protozoa processes very much resembling those of karyokinesis have already been observed taking place in the nucleus preparatory to its division. And although such processes do not present quite the same appearances as are to be met with in egg-cells, neither do the karyokinetic processes in tissue-cells, which in their sundry kinds exhibit great variations in this respect. Moreover, even if such were not the case, the bare fact that nuclear division is not invariably of the simple or direct character in the case of all Protozoa, is sufficient to show that the distinction now before us--like the one last dealt with--is by no means absolute. As in the case of sexual propagation, so in that of karyokinesis, processes which are common to all the Metazoa are not wholly without their foreshadowings in the Protozoa. And seeing how greatly exalted is the office of egg-cells--and even of tissue-cells--as compared with that of their supposed ancestry in protozoal cells, it seems to me scarcely to be wondered at if their specializations of function should be associated with corresponding peculiarities of structure--a general fact which would in no way militate against the doctrine of evolution. Could we know the whole truth, we should probably find that in order to endow the most primitive of egg-cells with its powers of marshalling its products into a living army of cell-battalions, such an egg-cell must have been passed through a course of developmental specialization of so elaborate a kind, that even the complex processes of karyokinesis are but a very inadequate expression thereof.

Probably I have now said enough to show that, remarkable and altogether exceptional as the properties of germ-cells of the multicellular organisms unquestionably show themselves to be, yet when these properties are traced back to their simplest beginnings in the unicellular organisms, they may fairly be regarded as fundamentally identical with the properties of living cells in general. Thus viewed, no line of real demarcation can be drawn between growth and reproduction, even of the sexual kind. The one process is, so to speak, physiologically continuous with the other; and hence, so far as the pre-embryonic stage of life-history is concerned, the facts cannot fairly be regarded as out of keeping with the theory of evolution.

I will now pass on to consider the embryogeny of the Metazoa, beginning at its earliest stage in the fertilization of the ovum. And here it is that the constructive argument in favour of evolution which is derived from embryology may be said properly to commence. For it is surely in itself a most suggestive fact that all the Metazoa begin their life in the same way, or under the same form and conditions. Omne vivum ex ovo. This is a formula which has now been found to apply throughout the whole range of the multicellular organisms. And seeing, as we have just seen, that the ovum is everywhere a single cell, the formula amounts to saying that, physiologically speaking, every Metazoa begins its life as a Protozoa, and every Metaphyton as a Protophyton[10].

[10] Even when propagated by budding, a multicellular organism has been ultimately derived from a germ-cell.

Now, if the theory of evolution is true, what should we expect to happen when these germ-cells are fertilized, and so enter upon their severally distinct processes of development? Assuredly we should expect to find that the higher organisms pass through the same phases of development as the lower organisms, up to the time when their higher characters begin to become apparent. If in the life-history of species these higher characters were gained by gradual improvement upon lower characters, and if the development of the higher individual is now a general recapitulation of that of its ancestral species, in studying this recapitulation we should expect to find the higher organism successively unfolding its higher characters from the lower ones through which its ancestral species had previously passed. And this is just what we do find. Take, for example, the case of the highest organism, Man. Like that of all other organisms, unicellular or multicellular, his development starts from the nucleus of a single cell. Again, like that of all the Metazoa and Metaphyta, his development starts from the specially elaborated nucleus of an egg-cell, or a nucleus which has been formed by the fusion of a male with a female element[11]. When his animality becomes established, he exhibits the fundamental anatomical qualities which characterize such lowly animals as polyps and jelly-fish. And even when he is marked off as a Vertebrate, it cannot be said whether he is to be a fish, a reptile, a bird, or a beast. Later on it becomes evident that he is to be a Mammal; but not till later still can it be said to which order of mammals he belongs.

[11] It has already been stated that both parthenogenesis and gemmation

are ultimately derived from sexual reproduction. It may now be added, on the other hand, that the earlier stages of parthenogenesis have been observed to occur sporadically in all sub-kingdoms of the Metaxoa, including the Vertebrata, and even the highest class, Mammalia. These earlier stages consist in spontaneous segmentations of the ovum; so that even if a virgin has ever conceived and borne a son, and even if such a fact in the human species has been unique, still it would not betoken any breach of physiological continuity. Indeed, according to Weismann's not improbable hypothesis touching the physiological meaning of polar bodies, such a fact need betoken nothing more than a slight disturbance of the complex machinery of ovulation, on account of which the ovum failed to eliminate from its substance an almost inconceivably minute portion of its nucleus.

Here, however, we must guard against an error which is frequently met with in popular expositions of this subject. It is not true that the embryonic phases in the development of a higher form always resemble so many adult stages of lower forms. This may or may not be the case; but what always is the case is, that the embryonic phases of the higher form resemble the corresponding phases of the lower forms. Thus, for example, it would be wrong to suppose that at any stage of his development a man resembles a jelly-fish. What he does resemble at an early stage of his development is the essential or groundplan of the jelly-fish, which that animal presents in its embryonic condition, or before it begins to assume its more specialized characters fitting it for its own particular sphere of life. The similarities, therefore, which it is the function of comparative embryology to reveal are the similarities of type or morphological plan: not similarities of specific detail. Specific details may have been added to this, that, and the other species for their own special requirements, after they had severally branched off from the common ancestral stem; and so could not be expected to recur in the life-history of an independent specific branch. The comparison therefore must be a comparison of embryo with embryo; not of embryos with adult forms.

* * * * *

In order to give a general idea of the results thus far yielded by a study of comparative embryology in the present connexion, I will devote the rest of this chapter to giving an outline sketch of the most important and best established of these results.

Histologically the ovum, or egg-cell, is nearly identical in all animals, whether vertebrate or invertebrate. Considered as a cell it is of large size, but actually it is not more than 1/100, and may be less than 1/200 of an inch in diameter. In man, as in most mammals, it is about 1/120. It is a more or less spherical body, presenting a thin transparent envelope, called the zona pellucida, which contains--first, the protoplasmic cell-substance or "yolk," within which lies, second, the nucleus or germinal vesicle, within which again lies, third, the nucleolus or germinal spot. This description is true of the egg-cells of all animals, if we add that in the case of the lowest animals--such as sponges, &c.--there is no enveloping membrane: the egg-cell is here a naked cell, and its constituent protoplasm, being thus unconfined, is free to perform protoplasmic movements, which it does after the manner, and with all the activity, of an amoeba. But even with respect to this matter of an enveloping membrane, there is no essential difference between an ovum of the lowest and an ovum of the highest animals. For in their early stages of development within the ovary the ova of the highest animals are likewise in the condition of naked cells, exhibiting amoebiform movements; the enveloping membrane of an ovum being the product of a later development. Moreover this membrane, when present, is usually provided with one or more minute apertures, through which the spermatozoa passes when fertilizing the ovum. It is remarkable that the spermatozoa know, so to speak, of the existence of these gate-ways,--their snake-like movements being directed towards them, presumably by a stimulus due to some emanation therefrom[12]. In the mammalian ovum, however, these apertures are exceedingly minute, and distributed all round the circumference of the pellucid envelope, as represented in this illustration (Fig. 32).

[12] The spermatozooids of certain plants can be strongly attracted towards a pipette which is filled with malic acid--crowding around and into it with avidity.

In thus saying that the ova of all animals are, so far as microscopes can reveal, substantially similar, I am of course speaking of the egg-cell proper, and not of what is popularly known as the egg. The egg of a bird, for example, is the egg-cell, plus an enormous aggregation of nutritive material, an egg-shell, and sundry other structures suited to the subsequent development of the egg-cell when separated from the parent's body. But all these accessories are, from our present point of view, accidental or adventitious. What we have now to understand by the ovum, the egg, or the egg-cell, is the microscopical germ which I have just described. So far then as this germ is concerned, we find that all multicellular organisms begin their existence in the same kind of structure, and that this structure is anatomically indistinguishable from that of the permanent form presented by the lowest, or unicellular organisms. But although anatomically indistinguishable, physiologically they present the sundry peculiarities already mentioned.

Now I have endeavoured to show that none of these peculiarities are such as to exclude--or even so much as to invalidate--the supposition of developmental continuity between the lowest egg-cells and the highest protozoal cells. It remains to show in this place, and on the other hand, that there is no breach of continuity between the lowest and the highest egg-cells; but, on the contrary, that the remarkable uniformity of the complex processes whereby their peculiar characters are exhibited to the histologist, is such as of itself to sustain the doctrine of continuity in a singularly forcible manner. On this account, therefore, and also because the facts will again have to be considered in another connexion when we come to deal with Weismann's theory of heredity, I will here briefly describe the processes in question.

We have already seen that the young egg-cell multiplies itself by simple binary division, after the manner of unicellular organisms in general--thereby indicating, as also by its amoebiform movements, its fundamental identity with such organisms in kind. But, as we have likewise seen, when the ovum ceases to resemble these organisms, by taking on its higher degree of functional capacity, it is no longer able to multiply itself in this manner. On the contrary, its cell-divisions are now of an endogenous character, and result in the formation of many different kinds of cells, in the order required for constructing the multicellular organism to which the whole series of processes eventually give rise. We have now to consider these processes seriatim.

First of all the nucleus discharges its polar bodies, as previously mentioned, and in the manner here depicted on the previous page. (Fig. 33.) It will be observed that the nucleus of the ovum, or the germinal vesicle as it is called, gets rid first of one and afterwards of the other polar body by an "indirect," or karyokinetic, process of division. (Fig. 33.) Extrusion of these bodies from the ovum (or it may be only from the nucleus) having been accomplished, what remains of the nucleus retires from the circumference of the ovum, and is called the female pronucleus. (Fig. 33. f. pn.) The ovum is now ready for fertilization. A similar emission of nuclear substance is said by some good observers to take place also from the male germ-cell, or spermatozoa, at or about the close of its development. The theories to which these facts have given rise will be considered in future chapters on Heredity.

Turning now to the mechanism of fertilization, the diagrams (Figs. 34, 35) represent what happens in the case of star-fish.

The sperm-cell, or spermatozoa, is seen in the act of penetrating the ovum. In the first figure it has already pierced the mucilaginous coat of the ovum, the limit of which is represented by a line through which the tail of the spermatozoa is passing: the head of the spermatozoa is just entering the ovum proper. It may be noted that, in the case of many animals, the general protoplasm of the ovum becomes aware, so to speak, of the approach of a spermatozoa, and sends up a process to meet it. (Fig. 35, A, B, C.) Several--or even many--spermatozoa may thus enter the coat of the ovum; but normally only one proceeds further, or right into the substance of the ovum, for the purpose of effecting fertilization. This spermatozoa, as soon as it enters the periphery of the yolk, or cell-substance proper, sets up a series of remarkable phenomena. First, its own head rapidly increases in size, and takes on the appearance of a cell-nucleus: this is called the male pronucleus. At the same time its tail begins to disappear, and the enlarged head proceeds to make its way directly towards the nucleus of the ovum which, as before stated, is now called the female pronucleus. The latter in its turn moves towards the former, and when the two meet they fuse into one mass, forming a new nucleus. Before the two actually meet, the spermatozoa has lost its tail altogether; and it is noteworthy that during its passage through the protoplasmic cell-contents of the ovum, it appears to exercise upon this protoplasm an attractive influence; for the granules of the latter in its vicinity dispose themselves around it in radiating lines. All these various phenomena are depicted in the above wood-cuts. (Figs. 34, 35.)

Fertilization having been thus effected by fusion of the male and female pronuclei into a single (or new) nucleus, this latter body proceeds to exhibit complicated processes of karyokinesis, which, as before shown, are preliminary to nuclear division in the case of egg-cells. Indeed the karyokinetic process may begin in both the pronuclei before their junction is effected; and, even when their junction is effected, it does not appear that complete fusion of the so-called chromatin elements of the two pronuclei takes place. For the purpose of explaining what this means, and still more for the purpose of giving a general idea of the karyokinetic processes as a whole, I will quote the following description of them, because, for terseness combined with lucidity, it is unsurpassable.

Researches, chiefly due to Flemming, have shown that the nucleus in very many tissues of higher plants and animals consists of a capsule containing a plasma of "achromatin," not deeply stained by re-agents, ramifying in which is a reticulum of "chromatin" consisting of fibres which readily take a deep stain. (Fig. 36, A). Further it is demonstrated that, when the cell is about to divide into two, definite and very remarkable movements take place in the nucleus, resulting in the disappearance of the capsule and in the arrangement of its fibres first in the form of a wreath (D), and subsequently (by the breaking of the loops formed by the fibres) in the form of a star (E). A further movement within the nucleus leads to an arrangement of the broken loops in two groups (F), the position of the open ends of the broken loops being reversed as compared with what previously obtained. Now the two groups diverge, and in many cases a striated appearance of the achromatin substance between the two groups of chromatin loops is observable (H). In some cases (especially egg-cells) this striated arrangement of the achromatin is then termed a "nucleus-spindle," and the group of chromatin loops (G, a) is known as "the equatorial plate." At each end of the nucleus-spindle in these cases there is often seen a star consisting of granules belonging to the general protoplasm of the cell (G, c). These are known as "polar stars." After the separation of the two sets of loops (H) the protoplasm of the general substance of the cell becomes constricted, and division occurs, so as to include a group of chromatin loops in each of the two fission products. Each of these then rearranges itself together with the associated chromatin into a nucleus such as was present in the mother cell to commence with (I)[13].

[13] Ray Lankester, Encyclop. Brit., 9th ed., Vol. XIX, pp. 832-3.

Since the above was published, however, further progress has been made.

In particular it has been found that the chromatin fibres pass from phase D to phase F by a process of longitudinal splitting (Fig. 37 g, h; Fig. 38, VI, VII)--which is a point of great importance for Weismann's theory of heredity,--and that the protoplasm outside the nucleus seems to take as important a part in the karyokinetic process as does the nuclear substance. For the so-called "attraction-spheres" (Fig. 38 II a, III, III a, VIII to XII), which were at first supposed to be of subordinate importance in the process as a whole, are now known to take an exceedingly active part in it (see especially IX to XI). Lastly, it may be added that there is a growing consensus of authoritative opinion, that the chromatin fibres are the seats of the material of heredity, or, in other words, that they contain those essential elements of the cell which endow the daughter-cells with their distinctive characters. Therefore, where the parent-cell is an ovum, it follows from this view that all hereditary qualities of the future organism are potentially present in the ultra-microscopical structure of the chromatin fibres.

VI., VII. The V-shaped filaments are splitting longitudinally; their structure of fine granules of chromatin is apparent in VII., which is more highly magnified. The conjugation of the pronuclei is apparently complete in VII. The attraction-spheres and achromatic spindle, although present, are not depicted in IV., V., VI., and VII.

VIII. Equatorial arrangement of the four chromatin loops in the middle of the now segmenting ovum: the achromatic substance forming a spindle-shaped system of granules with fibres radiating from the poles of the spindle (attraction-spheres); the chromatin forms an equatorial plate. (Compare Fig. 36 G.)

IX. Shows diagrammatically the commencing separation of the chromatin fibres of the conjugated nuclei, and the system of fibres radiating from the attraction-spheres. (Compare again Fig. 36 G.) p.c., polar circle; e.c., equatorial circle; c.c., central particle.

X. Further separation of the chromatin filaments. Each of the central particles of the attraction-spheres has divided into two.

XI. The chromatin fibres are becoming developed into the skeins of the two daughter-nuclei. These are still united by fibres of achromatin. The general protoplasm of the ovum is becoming divided.

XII. The two daughter-nuclei exhibit a chromatin network. Each of the attraction-spheres has divided into two, which are joined by fibres of achromatin, and connected with the periphery of the cell in the same way as in the original or parent sphere, III.]

As I shall have more to say about these processes in the next volume, when we shall see the important part which they bear in Weismann's theory of heredity, it is with a double purpose that I here introduce these yet further illustrations of them upon a somewhat larger scale. The present purpose is merely that of showing, more clearly than hitherto, the great complexity of these processes on the one hand, and, on the other, the general similarity which they display in egg-cells and in tissue-cells. But as in relation to this purpose the illustrations speak for themselves, I may now pass on at once to the history of embryonic development, which follows fertilization of the ovum.

* * * * *

We have seen that when the new nucleus of the fertilized ovum (which is formed by a coalescence of the male pronucleus with the female) has completed its karyokinetic processes, it is divided into two equal parts; that these are disposed at opposite poles of the ovum; and that the whole contents of the ovum are thereupon likewise divided into two equal parts, with the result that there are now two nucleated cells within the spherical wall of the ovum where before there had only been one. Moreover, we have also seen that a precisely similar series of events repeat themselves in each of these two cells, thus giving rise to four cells.

All this, it will be noticed, is a case of cell-multiplication, which differs from that which takes place in the unicellular organisms only in its being invariably preceded (as far as we know) by karyokinesis, and in the resulting cells being all confined within a common envelope, and so in not being free to separate. Nevertheless, from what has already been said, it will also be noticed that this feature makes all the difference between a Metazoa and a Protozoa; so that already the ovum presents the distinguishing character of a Metazoa.

I have dealt thus at considerable length upon the processes whereby the originally unicellular ovum and spermatozoa become converted into the multicellular germ, because I do not know of any other exposition of the argument from Embryology where this, the first stage of the argument, has been adequately treated. Yet it is evident that the fact of all the processes above described being so similar in the case of sexual (or metazoal) reproduction among the innumerable organisms where it occurs, constitutes in itself a strong argument in favour of evolution. For the mechanism of fertilization, and all the processes which even thus far we have seen to follow therefrom, are hereby shown to be not only highly complex, but likewise highly specialized. Therefore, the remarkable similarity which they present throughout the whole animal kingdom--not to speak of the vegetable--is expressive of organic continuity, rather than of absolute discontinuity in every case, as the theory of special creation must necessarily suppose. And it is evident that this argument is strong in proportion to the uniformity, the specialization, and the complexity of the processes in question.

Having occupied so much space with supplying what appear to me the deficiencies in previous expositions of the argument from Embryology, I can now afford to take only a very general view of the more important features of this argument as they are successively furnished by all the later stages of individual development. But this is of little consequence, seeing that from the point at which we have now arrived previous expositions of the argument are both good and numerous. The following then is to be regarded as a mere sketch of the evidences of phyletic (or ancestral) evolution, which are so abundantly furnished by all the subsequent phases of ontogenetic (or individual) evolution.

The multicellular body which is formed by the series of segmentations above described is at first a sphere of cells (Fig. 40). Soon, however, a watery fluid gathers in the centre, and progressively pushes the cells towards the circumference, until they there constitute a single layer. The ovum, therefore, is now in the form of a hollow sphere containing fluid, confined within a continuous wall of cells (Fig. 41 A). The next thing that happens is a pitting in of one portion of the sphere (B). The pit becomes deeper and deeper, until there is a complete invagination of this part of the sphere--the cells which constitute it being progressively pushed inwards until they come into contact with those at the opposite pole of the ovum. Consequently, instead of a hollow sphere of cells, the ovum now becomes an open sac, the walls of which are composed of a double layer of cells (C). The ovum is now what has been called a gastrula; and it is of importance to observe that probably all the Metazoa pass through this stage. At any rate it has been found to occur in all the main divisions of the animal kingdom, as a glance at the accompanying figures will serve to show (Fig. 42)[14]. Moreover many of the lower kinds of Metazoa never pass beyond it; but are all their lives nothing else than gastrul? wherein the orifice becomes the mouth of the animal, the internal or invaginated layer of cells the stomach, and the outer layer the skin. So that if we take a child's india-rubber ball, of the hollow kind with a hole in it, and push in one side with our fingers till internal contact is established all round, by then holding the indented side downwards we should get a very fair anatomical model of a gastric form, such as is presented by the adult condition of many of the most primitive Metazoa--especially the lower Coelenterata. The preceding figures represent two other such forms in nature, the first locomotive and transitory, the second fixed and permanent (Figs. 43, 44).

[14] In most vertebrated animals this process of gastrulation has been more or less superseded by another, which is called delamination; but it scarcely seems necessary for our present purposes to describe the latter. For not only does it eventually lead to the same result as gastrulation--i. e. the converting of the ovum into a double-walled sac,--but there is good evidence among the lower Vertebrata of its being preceded by gastrulation; so that, even as to the higher Vertebrata, embryologists are pretty well agreed that delamination has been but a later development of, or possibly improvement upon, gastrulation.

Here, then, we leave the lower forms of Metazoa in their condition of permanent gastrul? They differ from the transitory stage of other Metazoa only in being enormously larger (owing to greatly further growth, without any further development as to matters of fundamental importance), and in having sundry tentacles and other organs added later on to meet their special requirements. The point to remember is, that in all cases a gastrula is an open sac composed of two layers of cells--the outer layer being called the ectoderm, and the inner the endoderm. They have also been called the animal layer and the vegetative layer, because it is the outer layer (ectoderm) that gives rise to all the organs of sensation and movement--viz. the skin, the nervous system, and the muscular system; while it is the inner layer (endoderm) that gives rise to all the organs of nutrition and reproduction. It is desirable only further to explain that gastrulation does not take place in all the Metazoa after exactly the same plan. In different lines of descent various and often considerable modifications of the original and most simple plan have been introduced; but I will not burden the present exposition by describing these modifications[15]. It is enough for us that they always end in the formation of the two primary layers of ectoderm and endoderm.

[15] The most extreme of them is that which is mentioned in the last foot-note.

The next stage of differentiation is common to all the Metazoa, except those lowest forms which, as we have just seen, remain permanently as large gastrul? with sundry specialized additions in the way of tentacles, &c. This stage of differentiation consists in the formation of either a pouch or an additional layer between the ectoderm and the endoderm, which is called the mesoderm. It is probably in most cases derived from the endoderm, but the exact mode of its derivation is still somewhat obscure. Sometimes it has the appearance of itself constituting two layers; but it is needless to go into these details; for in any case the ultimate result is the same--viz. that of converting the Metazoa into the form of a tube, the walls of which are composed of concentric layers of cells. The outermost layer afterwards gives rise to the epidermis with its various appendages, and also to the central nervous system with its organs of special sense. The median layer gives rise to the voluntary muscles, bones, cartilages, &c., the nutritive systems of the blood, the chyle, the lymph, and the muscular tube of the intestine. Lastly, the innermost layer developes into the epithelium lining of the intestine, with its various appendages of liver, lungs, intestinal glands, &c.

I have just said that this three or four layered stage is shared by all the Metazoa, except those very lowest forms--such as sponges and jelly-fish--which do not pass on to it. But from this point the developmental histories of all the main branches of the Metazoa diverge--the Vermes, the Echinodermata, the Mollusca, the Articulata, and the Vertebrata, each taking a different road in their subsequent evolution. I will therefore confine attention to only one of these several roads or methods, namely, that which is followed by the Vertebrata--observing merely that, if space permitted, the same principles of progressive though diverging histories of evolution would equally well admit of being traced in all the other sub-kingdoms which have just been named.

In order to trace these principles in the case of the Vertebrata, it is desirable first of all to obtain an idea of the anatomical features which most essentially distinguish the sub-kingdom as a whole. The following, then, is what may be termed the ideal plan of vertebrate organization, as given by Prof. Heckel. First, occupying the major axis of body we perceive the primitive vertebral column. The parts lying above this axis are those which have been developed from the ectoderm and mesoderm--viz. voluntary muscles, central nervous system, and organs of special sense. The parts lying below this axis are for the most part those which have been developed from the endoderm--namely, the digestive tract with its glandular appendages, the circulating system and the respiratory system. In transverse section, therefore, the ideal vertebrate consists of a solid axis, with a small tube occupied by the nervous system above, and a large tube, or body-cavity, below. This body-cavity contains the viscera, breathing organs, and heart, with its prolongations into the main blood-vessels of the organism. Lastly, on either side of the central axis are to be found large masses of muscle--two on the dorsal and two on the ventral. As yet, however, there are no limbs, nor even any bony skeleton, for the primitive vertebral column is hitherto unossified cartilage. This ideal animal, therefore, is to all appearance as much like a worm as a fish, and swims by means of a lateral undulation of its whole body, assisted, perhaps, by a dorsal fin formed out of skin.

Now I should not have presented this ideal representation of a primitive vertebrate--for I have very little faith in the "scientific use of the imagination" where it aspires to discharge the functions of a Creator in the manufacture of archetypal forms--I say I should not have presented this ideal representative of a primitive vertebrate, were it not that the ideal is actually realized in a still existing animal. For there still survives what must be an immensely archaic form of vertebrate, whose anatomy is almost identical with that of the imaginary type which has just been described. I allude, of course, to Amphioxus, which is by far the most primitive or generalized type of vertebrated animal hitherto discovered. Indeed, we may say that this remarkable creature is almost as nearly allied to a worm as it is to a fish. For it has no specialized head, and therefore no skull, brain, or jaws: it is destitute alike of limbs, of a centralized heart, of developed liver, kidneys, and, in short, of most of the organs which belong to the other Vertebrata. It presents, however, a rudimentary backbone, in the form of what is called a notochord. Now a primitive dorsal axis of this kind occurs at a very early period of embryonic life in all vertebrated animals; but, with the exception of Amphioxus, in all other existing Vertebrata this structure is not itself destined to become the permanent or bony vertebral column. On the contrary, it gives way to, or is replaced by, this permanent bony structure at a later stage of development. Consequently, it is very suggestive that so distinctively embryonic a structure as this temporary cartilaginous axis of all the other known Vertebrata should be found actually persisting to the present day as the permanent axis of Amphioxus. In many other respects, likewise, the early embryonic history of other Vertebrata refers us to the permanent condition of Amphioxus. In particular, we must notice that the wall of the neck is always perforated by what in Amphioxus are the gill-openings, and that the blood-vessels as they proceed from the heart are always distributed in the form of what are called gill-arches, adapted to convey the blood round or through the gills for the purpose of aeration. In all existing fish and other gill-breathing Vertebrata, this arrangement is permanent. It is likewise met with in a peculiar kind of worm, called Balanoglossus--a creature so peculiar, indeed, that it has been constituted by Gegenbaur a class all by itself. We can see by the wood-cuts that it presents a series of gill-slits, like the homologous parts of the fishes with which it is compared--i. e. fishes of a comparatively low type of organization, which dates from a time before the development of external gills. (Figs. 48, 49, 50.) Now, as I have already said, these gill-slits are supported internally by the gill-arches, or the blood-vessels which convey the blood to be oxygenized in the branchial apparatus (see below, Figs. 51, 52, 53); and the whole arrangement is developed from the anterior part of the intestine--as is likewise the respiratory mechanism of all the gill-breathing Vertebrata. That so close a parallel to this peculiar mechanism should be met with in a worm, is a strong additional piece of evidence pointing to the derivation of the Vertebrata from the Vermes.

Well, I have just said that in all the gill-breathing Vertebrata, this mechanism of gill-slits and vascular gill-arches in the front part of the intestinal tract is permanent. But in the air-breathing Vertebrata such an arrangement would obviously be of no use. Consequently, the gill-slits in the sides of the neck (see Figs. 16 and 57, 58), and the gill-arches of the large blood-vessels (Figs.

54, 55, 56), are here exhibited only as transitory phases of development. But as such they occur in all air-breathing Vertebrata. And, as if to make the homologies as striking as possible, at the time when the gill-slits and the gill-arches are developed in the embryonic young of air-breathing Vertebrata, the heart is constructed upon the fish-like type. That is to say, it is placed far forwards, and, from having been a simple tube as in Worms, is now divided into two chambers, as in Fish. Later on it becomes progressively pushed further back between the developing lungs, while it progressively acquires the three cavities distinctive of Amphibia, and finally the four cavities belonging only to the complete double circulation of Birds and Mammals. Moreover, it has now been satisfactorily shown that the lungs of air-breathing Vertebrata, which are thus destined to supersede the function of gills, are themselves the modified swim-bladder or float, which belongs to Fish. Consequently, all these progressive modifications in the important organs of circulation and respiration in the air-breathing Vertebrata, together make up as complete a history of their aquatic pedigree as it would be possible for the most exacting critic to require.

If space permitted, it would be easy to present abundance of additional evidence to the same effect from the development of the skeleton, the skull, the brain, the sense-organs, and, in short, of every constituent part of the vertebrate organization. Even without any anatomical dissection, the similarity of all vertebrated embryos at comparable stages of development admits of being strikingly shown, if we merely place the embryos one beside the other. Here, for instance, are the embryos of a fish, a salamander, a tortoise, a bird, and four different mammals. In each case three comparable stages of development are represented. Now, if we read the series horizontally, we can see that there is very little difference between the eight animals at the earliest of the three stages represented--all having fish-like tails, gill-slits, and so on. In the next stage further differentiation has taken place, but it will be observed that the limbs are still so rudimentary that even in the case of Man they are considerably shorter than the tail. But in the third stage the distinctive characters are well marked.

* * * * *

So much then for an outline sketch of the main features in the embryonic history of the Vertebrata. But it must be remembered that the science of comparative embryology extends to each of the other three great branches of the tree of life, where these take their origin, through the worms, from the still lower, or gastric, forms. And in each of these three great branches-- namely, the Echinodermata, the Mollusca, and the Arthropoda--we have a repetition of just the same kind of evidence in favour of continuous descent, with adaptive modification in sundry lines, as that which I have thus briefly sketched in the case of the Vertebrata. The roads are different, but the method of travelling is the same. Moreover, when the embryology of the Worms is closely studied, the origin of these different roads admits of being clearly traced. So that when all this mass of evidence is taken together, we cannot wonder that evolutionists should now regard the science of comparative embryology as the principal witness to their theory.

CHAPTER V.

PALEONTOLOGY.

The present Chapter will be devoted to a consideration of the evidence of organic evolution which has been furnished by the researches of geologists. On account of its direct or historical nature, this branch of evidence is popularly regarded as the most important--so much so, indeed, that in the opinion of most educated persons the whole doctrine of organic evolution must stand or fall according to the so-called "testimony of the rocks." Now, without at all denying the peculiar importance of this line of evidence, I must begin by remarking that it does not present the denominating importance which popular judgment assigns to it. For although popular judgment is right in regarding the testimony of the rocks as of the nature of a history, this judgment, as a rule, is very inadequately acquainted with the great imperfections of that history. Knowing in a general way what magnificent advances the science of geology has made during the present century, the public mind is more or less imbued with the notion, that because we now possess a tolerably complete record of the chronological succession of geological formations, we must therefore possess a correspondingly complete record of the chronological succession of the forms of life which from time to time have peopled the globe. Now in one sense this notion is partly true, but in another sense it is profoundly false. It is partly true if we have regard only to those larger divisions of the vegetable or animal kingdoms which naturalists designate by the terms classes and orders. But the notion becomes progressively more untrue when it is applied to families and genera, while it is most of all untrue when applied to species. That this must be so may be rendered apparent by two considerations.

In the first place, it does not follow that because we have a tolerably complete record of the succession of geological formations, we have therefore any correspondingly complete record of their fossiliferous contents. The work of determining the relative ages of the rocks does not require that every cubic mile of the earth's surface should be separately examined, in order to find all the different fossils which it may contain. Were this the case, we should hitherto have made but very small progress in our reading of the testimony of the rocks. The relative ages of the rocks are determined by broad comparative surveys over extensive areas; and although the identification of widely separated deposits is often greatly assisted by a study of their fossiliferous contents, the mere pricking of a continent here and there is all that is required for this purpose. Hence, the accuracy of our information touching the relative ages of geological strata does not depend upon--and, therefore, does not betoken--any equivalent accuracy of knowledge touching the fossiliferous material which these strata may at the present time actually contain. And, as we well know, the opportunities which the geologist has of discovering fossils are extremely limited, if we consider these opportunities in relation to the area of geological formations. The larger portion of the earth's surface is buried beneath the sea; and much the larger portion of the fossiliferous deposits on shore are no less hopelessly buried beneath the land. Therefore it is only upon the fractional portion of the earth's surface which at the present time happens to be actually exposed to his view that the geologist is able to prosecute his search for fossils. But even here how miserably inadequate this search has hitherto been! With the exception of a scratch or two in the continents of Asia and America, together with a somewhat larger number of similar scratches over the continent of Europe, even that comparatively small portion of the earth's surface which is available for the purpose has been hitherto quite unexplored by the paleontologist. How enormously rich a store of material remains to be unearthed by the future scratchings of this surface, we may dimly surmise from the astonishing world of bygone life which is now being revealed in the newly discovered fossiliferous deposits on the continent of America.

But, besides all this, we must remember, in the second place, that all the fossiliferous deposits in the world, even if they could be thoroughly explored, would still prove highly imperfect, considered as a history of extinct forms of life. In order that many of these forms should have been preserved as fossils, it is necessary that they should have died upon a surface neither too hard nor too soft to admit of their leaving an impression; that this surface should afterwards have hardened sufficiently to retain the impression; that it should then have been protected from the erosion of water, as well as from the disintegrating influence of the air; and yet that it should not have sunk far enough beneath the surface to have come within the no less disintegrating influence of subterranean heat. Remembering thus, as a general rule, how many conditions require to have met before a fossil can have been both formed and preserved, we must conclude that the geological record is probably as imperfect in itself as are our opportunities of reading even the little that has been recorded. If we speak of it as a history of the succession of life upon the planet, we must allow, on the one hand, that it is a history which merits the name of a "chapter of accidents"; and, on the other hand, that during the whole course of its compilation pages were being destroyed as fast as others were being formed, while even of those that remain it is only a word, a line, or at most a short paragraph here and there, that we are permitted to see. With so fragmentary a record as this to study, I do not think it is too much to say that no conclusions can be fairly based upon it, merely from the absence of testimony. Only if the testimony were positively opposed to the theory of descent, could any argument be fairly raised against that theory on the grounds of this testimony. In other words, if any of the fossils hitherto discovered prove the order of succession to have been incompatible with the theory of genetic descent, then the record may fairly be adduced in argument, because we should then be in possession of definite information of a positive kind, instead of a mere absence of information of any kind. But if the adverse argument reaches only to the extent of maintaining that the geological record does not furnish us with so complete a series of "connecting links" as we might have expected, then, I think, the argument is futile. Even in the case of human histories, written with the intentional purpose of conveying information, it is an unsafe thing to infer the non-occurrence of an event from a mere silence of the historian--and this especially in matters of comparatively small detail, such as would correspond (in the present analogy) to the occurrence of species and genera as connecting links. And, of course, if the history had only come down to us in fragments, no one would attach any importance at all to what might have been only the apparent silence of the historian.

In view, then, of the unfortunate imperfection of the geological record per se, as well as of the no less unfortunate limitation of our means of reading even so much of the record as has come down to us, I conclude that this record can only be fairly used in two ways. It may fairly be examined for positive testimony against the theory of descent, or for proof of the presence of organic remains of a high order of development in a low level of strata. And it may be fairly examined for negative testimony, or for the absence of connecting links, if the search be confined to the larger taxonomic divisions of the fauna and flora of the world. The more minute these divisions, the more restricted must have been the areas of their origin, and hence the less likelihood of their having been preserved in the fossil state, or of our finding them even if they have been. Therefore, if the theory of evolution is true, we ought not to expect from the geological record a full history of specific changes in any but at most a comparatively small number of instances, where local circumstances happen to have been favourable for the writing and preservation of such a history. But we might reasonably expect to find a general concurrence of geological testimony to the larger fact--namely, of there having been throughout all geological time a uniform progression as regards the larger taxonomic divisions. And, as I will next proceed to show, this is, in a general way, what we do find, although not altogether without some important exceptions, with which I shall deal in an Appendix.

There is no positive proof against the theory of descent to be drawn from a study of Paleontology, or proof of the presence of any kind of fossils in strata where the fact of their presence is incompatible with the theory of evolution. On the other hand, there is an enormous body of uniform evidence to prove two general facts of the highest importance in the present connexion. The first of these general facts is, that an increase in the diversity of types both of plants and animals has been constant and progressive from the earliest to the latest times, as we should anticipate that it must have been on the theory of descent in ever-ramifying lines of pedigree. And the second general fact is, that through all these branching lines of ever-multiplying types, from the first appearance of each of them to their latest known conditions, there is overwhelming evidence of one great law of organic nature--the law of gradual advance from the general to the special, from the low to the high, from the simple to the complex.

Now, the importance of these large and general facts in the present connexion must be at once apparent; but it may perhaps be rendered more so if we try to imagine how the case would have stood supposing geological investigation to have yielded in this matter an opposite result, or even so much as an equivocal result. If it had yielded an opposite result, if the lower geological formations were found to contain as many, as diverse, and as highly organized types as the later geological formations, clearly there would have been no room at all for any theory of progressive evolution. And, by parity of reasoning, in whatever degree such a state of matters were found to prevail, in that degree would the theory in question have been discredited. But seeing that these opposite principles do not prevail in any (relatively speaking) considerable degree[16], we have so far positive testimony of the largest and most massive character in favour of this theory. For while all these large and general facts are very much what they ought to be according to this theory, they cannot be held to lend any support at all to the rival theory. In other words, it is clearly no essential part of the theory of special creation that species should everywhere exhibit this gradual multiplication as to number, coupled with a gradual diversification and general elevation of types, in all the growing branches of the tree of life. No one could adopt seriously the jocular lines of Burns, to the effect that the Creator required to practise his prentice hand on lower types before advancing to the formation of higher. Yet, without some such assumption, it would be impossible to explain, on the theory of independent creations, why there should have been this gradual advance from the few to the many, from the general to the special, from the low to the high.

I submit, then, that so far as the largest and most general principles in the matter of Paleontology are concerned, we have about as strong and massive a body of evidence as we could reasonably expect this branch of science to yield; for it is at once enormous in amount and positive in character. Therefore, if I do not further enlarge upon the evidence which we here have, as it were en masse, it is only because I do not feel that any words could add to its obvious significance. It may best be allowed to speak for itself in the millions of facts which are condensed in this tabular statement of the order of succession of all the known forms of animal life, as presented by the eminent paleontologist, Professor Cope[17].

[17] For difficulties and objections, see Appendix.

Or, taking a still more general survey, this tabular statement may be still further condensed, and presented in a diagrammatic form, as it has been by another eminent American paleontologist, Prof. Le Conte, in his excellent little treatise on Evolution and its Relations to Religious Thought. The following is his diagrammatic representation, with his remarks thereon.

When each ruling class declined in importance, it did not perish, but continued in a subordinate position. Thus, the whole organic kingdom became not only higher and higher in its highest forms, but also more and more complex in its structure and in the interaction of its correlated parts. The whole process and its result is roughly represented in the accompanying diagram, in which A B represents the course of geological time, and the curve, the rise, culmination, and decline of successive dominate classes.

I will here leave the evidence which is thus yielded by the most general principles that have been established by the science of Paleontology; and I will devote the rest of this chapter to a detailed consideration of a few highly special lines of evidence. By thus suddenly passing from one extreme to the other, I hope to convey the best idea that can be conveyed within a brief compass of the minuteness, as well as the extent, of the testimony which is furnished by the rocks.

* * * * *

When Darwin first published his Origin of Species, adverse critics fastened upon the "missing-link" argument as the strongest that they could bring against the theory of descent. Although Darwin had himself strongly insisted on the imperfection of the geological record, and the consequent precariousness of any negative conclusions raised upon it, these critics maintained that he was making too great a demand upon the argument from ignorance--that, even allowing for the imperfection of the record, they would certainly have expected at least a few cases of testimony to specific transmutation. For, they urged in effect, looking to the enormous profusion of the extinct species on the one hand, and to the immense number of known fossils on the other, it was incredible that no satisfactory instances of specific transmutation should ever have been brought to light, if such transmutation had ever occurred in the universal manner which the theory was bound to suppose. But since Darwin first published his great work paleontologists have been very active in discovering and exploring fossiliferous beds in sundry parts of the world; and the result of their labours has been to supply so many of the previously missing links that the voice of competent criticism in this matter has now been well-nigh silenced. Indeed, the material thus furnished to an advocate of evolution at the present time is so abundant that his principal difficulty is to select his samples. I think, however, that the most satisfactory result will be gained if I restrict my exposition to a minute account of some few series of connecting links, rather than if I were to take a more general survey of a larger number. I will, therefore, confine the survey to the animal kingdom, and there mention only some of the cases which have yielded well-detailed proof of continuous differentiation.

It is obvious that the parts of animals most likely to have been preserved in such a continuous series of fossils as the present line of evidence requires, would have been the hard parts. These are horns, bones, teeth, and shells. Therefore I will consider each of these four classes of structures separately.

* * * * *

Horns wherever they occur, are found to be of high importance for purposes of classification. They are restricted to the Ruminants, and appear under three different forms or types--namely solid, as in antelopes; hollow, as in sheep; and deciduous, as in deer. Now, in each of these divisions we have a tolerably complete paleontological history of the evolution of horns. The early ruminants were altogether hornless (Fig. 60). Then, in the middle Miocene, the first antelopes appeared with tiny horns, which progressively increased in size among the ever-multiplying species of antelopes until the present day. But it is in the deer tribe that we meet with even better evidence touching the progressive evolution of horns; because here not only size, but shape, is concerned. For deer's horns, or antlers, are arborescent; and hence in their case we have an opportunity of reading the history, not only of a progressive growth in size, but also of an increasing development of form. Among the older members of the tribe, in the lower Miocene, there are no horns at all. In the mid-Miocene we meet with two-pronged horns (Cervus dicrocerus, Figs. 61, 62, 1/5 nat. size). Next, in the upper Miocene (C. matheronis, Fig. 63, 1/8 nat. size), and extending into the Pliocene (C.

pardinensis, Fig. 64, 1/18 nat. size), we meet with three-pronged horns. Then, in the Pliocene we find also four-pronged horns (C. issiodorensis, Fig. 65, 1/16 nat. size), leading us to five-pronged (C. tetraceros). Lastly, in the Forest-bed of Norfolk we meet with arborescent horns (C. Sedgwickii, Fig. 66, 1/35 nat. size). The life-history of existing stags furnishes a parallel development (Fig. 67), beginning with a single horn (which has not yet been found paleontologically), going on to two prongs, three prongs, four prongs, and afterwards branching.

* * * * *

Coming now to bones, we have a singularly complete record of transition from one type or pattern of structure to another in the phylogenetic history of tails. This has been so clearly and so tersely conveyed by Prof. Le Conte, that I cannot do better than quote his statement.

It has long been noticed that there are among fishes two styles of tail-fins. These are the even-lobed, or homocercal (Fig. 68), and the uneven-lobed, or heterocercal (Fig. 69). The one is characteristic of ordinary fishes (teleosts), the other of sharks and some other orders. In structure the difference is even more fundamental than in form. In the former style the backbone stops abruptly in a series of short, enlarged joints, and thence sends off rays to form the tail-fin (Fig. 68); in the latter the backbone runs through the fin to its very point, growing slenderer by degrees, and giving off rays above and below from each joint, but the rays on the lower side are much longer (Fig. 69). This type of fin is, therefore, vertebrated, the other non-vertebrated. Figs. 68 and 69 show these two types in form and structure. But there is still another type found only in the lowest and most generalized forms of fishes. In these the tail-fin is vertebrated and yet symmetrical. This type is shown in Fig. 70.

Now, in the development of a teleost fish (Fig. 68), as has been shown by Alexander Agassiz, the tail-fin is first like Fig. 70; then becomes heterocercal, like Fig. 69; and, finally, becomes homocercal like Fig. 68. Why so? Not because there is any special advantage in this succession of forms; for the changes take place either in the egg or else in very early embryonic states. The answer is found in the fact that this is the order of change in the phylogenetic series. The earliest fish-tails were either like Fig. 69 or Fig. 70;

never like Fig. 68. The earliest of all were almost certainly like Fig. 70; then they became like Fig. 69; and, finally, only much later in geological history (Jurassic or Cretaceous), they became like Fig. 68. This order of change is still retained in the embryonic development of the last introduced and most specialized order of existing fishes. The family history is repeated in the individual history.

Similar changes have taken place in the form and structure of birds' tails. The earliest bird known--the Jurassic --had a long reptilian tail of twenty-one joints, each joint bearing a feather on each side, right and left (Fig. 71): [see also Fig. 73]. In the typical modern bird, on the contrary, the tail-joints are diminished in number, shortened up, and enlarged, and give out long feathers, fan-like, to form the so-called tail (Fig. 72). The Archeopteryx' tail is vertebrated, the typical bird's non-vertebrated. This shortening up of the tail did not take place at once, but gradually. The Cretaceous birds, intermediate in time, had tails intermediate in structure. The Hesperornis of Marsh had twelve joints. At first--in Jurassic strata--the tail is fully a half of the whole vertebral column. It then gradually shortens up until it becomes the aborted organ of typical modern birds. Now, in embryonic development, the tail of the modern typical bird passes through all these stages. At first the tail is nearly one half the whole vertebral column; then, as development goes on, while the rest of the body grows, the growth of the tail stops, and thus finally becomes the aborted organ we now find. The ontogeny still passes through the stages of the phylogeny. The same is true of all tailless animals.

he extinct Archeopteryx alluded to presents throughout its whole organization a most interesting assemblage of "generalized characters." For example, its teeth, and its still unreduced digits of the wings (which, like those of the feet, are covered with scales), refer us, with almost as much force as does the vertebrated tail, to the Sauropsidian type--or the trunk from which birds and reptiles have diverged.

We will next consider the paleontological evidence which we now possess of the evolution of mammalian limbs, with special reference to the hoofed animals, where this line of evidence happens to be most complete.

I may best begin by describing the bones as these occur in the sundry branches of the mammalian type now living. As we shall presently see, the modifications which the limbs have undergone in these sundry branches chiefly consist in the suppression of some parts and the exaggerated development of others. But, by comparing all mammalian limbs together, it is easy to obtain a generalized type of mammalian limb, which in actual life is perhaps most nearly conformed to in the case of bears. I will therefore choose the bear for the purpose of briefly expounding the bones of mammalian limbs in general--merely asking it to be understood, that although in the case of many other mammalia some of these bones may be dwindled or altogether absent, while others may be greatly exaggerated as to relative size, in no case do any additional bones appear.

On looking, then, at the skeleton of a bear (Fig. 74), the first thing to observe is that there is a perfect serial homology between the bones of the hind legs and of the fore legs. The thigh-bone, or femur, corresponds to the shoulder-bone, or humerus; the two shank bones (tibia and fibula) correspond to the two arm-bones (radius and ulna); the many little ankle-bones (tarsals) correspond to the many little wrist-bones (carpals); the foot-bones (meta-tarsals) correspond to the hand-bones (meta-carpals); and, lastly, the bones of each of the toes correspond to those of each of the fingers.

The next thing to observe is, that the disposition of bones in the case of the bear is such that the animal walks in the way that has been called plantigrade. That is to say, all the bones of the fingers, as well as those of the toes, feet, and ankles, rest upon the ground, or help to constitute the "soles." Our own feet are constructed on a closely similar pattern. But in the majority of living mammalian forms this is not the case. For the majority of mammals are what has been called digitigrade. That is to say, the bones of the limb are so disposed that both the foot and hand bones, and therefore also the ankle and wrist, are removed from the ground altogether, so that the animal walks exclusively upon its toes and fingers--as in the case of this skeleton (Fig. 75), which is the skeleton of a lion. The next figures display a series of limbs, showing the progressive passage of a completely plantigrade into a highly digitigrade type--the curved lines of connexion serving to indicate the homologous bones (Figs. 76, 77).

I will now proceed to detail the history of mammalian limbs, as this has been recorded for us in fossil remains.

The most generalized or primitive types of limb hitherto discovered in any vertebrated animal above the class of fishes, are those which are met with in some of the extinct aquatic reptiles. Here, for instance, is a diagram of the left hind limb of Baptanodon discus (Fig. 78). It has six rows of little symmetrical bones springing from a leg-like origin. But the whole structure resembles the fin of a fish about as nearly as it does the leg of a mammal. For not only are there six rows of bones, instead of five, suggestive of the numerous rays which characterise the fin of a fish; but the structure as a whole, having been covered over with blubber and skin, was throughout flexible and unjointed--thus in function, even more than in structure, resembling a fin. In this respect, also, it must have resembled the paddle of a whale (see Fig. 79); but of course the great difference will be noted, that the paddle of a whale reveals the dwindled though still clearly typical bones of a true mammalian limb; so that although in outward form and function these two paddles are alike, their inward structure clearly shows that while the one testifies to the absence of evolution, the other testifies to the presence of degeneration. If the paddle of Baptanodon had occurred in a whale, or the paddle of a whale had occurred in Baptanodon, either fact would in itself have been well-nigh destructive of the whole theory of evolution.

Such, then, is the most generalized as it is the most ancient type of vertebrate limb above the class of fishes. Obviously it is a type suited only to aquatic life. Consequently, when aquatic Vertebrata began to become terrestrial, the type would have needed modification in order to serve for terrestrial locomotion. In particular, it would have needed to gain in consolidation and in firmness, which means that it would have needed also to become jointed. Accordingly, we find that this archaic type gave place in land-reptiles to the exigencies of these requirements. Here for example is a diagram, copied from Gegenbaur, of the right fore-foot of Chelydra serpentina (Fig. 78). As compared with the homologous limb of its purely aquatic predecessor, there is to be noticed the disappearance of one of the six rows of small bones, a confluence of some of the remainder in the other five rows, a duplication of the arm-bone into a radius and ulna, in order to admit of jointed rotation of the hand, and a general disposition of the small bones below these arm-bones, which clearly foreshadows the joint of the wrist. Indeed, in this fore-foot of Chelydra, a child could trace all the principal homologies of the mammalian counterpart, growing, like the next stage in a dissolving view, out of the primitive paddle of Baptanodon--namely, first the radius and ulna, next the carpals, then the meta-carpals, and, lastly, the three phalanges in each of the five digits.

Such a type of foot no doubt admirably meets the requirements of slow reptilian locomotion over swampy ground. But for anything like rapid locomotion over hard and uneven ground, greater modifications would be needed. Such modifications, however, need not be other in kind: it is enough that they should continue in the same line of advance, so as to reach a higher degree of firmness, combined with better joints. Accordingly we find that this took place, not indeed among reptiles, whose habits of cold-blooded life have not changed, but among their warm-blooded descendants, the mammals. Moreover, when we examine the whole mammalian series, we find that the required modifications must have taken place in slightly different ways in three lines of descent simultaneously. We have first the plantigrade and digitigrade modifications already mentioned (pp. 178, 179) Of these the plantigrade walking entailed least change, because most resembling the ancestral or lizard-like mode of progression. All that was here needed was a general improvement as to relative lengths of bones, with greater consolidation and greater flexibility of joints. Therefore I need not say anything more about the plantigrade division. But the digitigrade modification necessitated a change of structural plan, to the extent of raising the wrist and ankle joints off the ground, so as to make the quadruped walk on its fingers and toes. We meet with an interesting case of this transition in the existing hare, which while at rest supports itself on the whole hind foot after the manner of a plantigrade animal, but when running does so upon the ends of its toes, after the manner of a digitigrade animal.

It is of importance for us to note that this transition from the original plantigrade to the more recent digitigrade type, has been carried out on two slightly different plans in two different lines of mammalian descent. The hoofed mammals--which are all digitigrade--are sub-classified as artiodactyls and perissodactyls, i. e. even-toed and odd-toed. Now, whether an animal has an even or an odd number of toes may seem a curiously artificial distinction on which to found so important a classification of the mammalian group. But if we look at the matter from a less empirical and more intelligent point of view, we shall see that the alternative of having an even or an odd number of toes carries with it alternative consequences of a practically important kind to any animal of the digitigrade type. For suppose an aboriginal five-toed animal, walking on the ends of its five toes, to be called upon to resign some of his toes. If he is left with an even number, it must be two or four; and in either case the animal would gain the firmest support by so disposing his toes as to admit of the axis of his foot passing between an equal number of them--whether it be one or two toes on each side. On the other hand, if our early mammal were called upon to retain an odd number of toes, he would gain best support by adjusting matters so that the axis of his foot should be coincident with his middle toe, whether this were his only toe, or whether he had one on either side of it. This consideration shows that the classification into even-toed and odd-toed is not so artificial as it no doubt at first sight appears. Let us, then, consider the stages in the evolution of both these types of feet.

Going back to the reptile Chelydra, it will be observed that the axis of the foot passes down the middle toe, which is therefore supported by two toes on either side (Fig. 78). It may also be noticed that the wrist or ankle bones do not interlock, either with one another or with the bones of the hand or foot below them. This, of course, would give a weak foot, suited to slow progression over marshy ground--which, as we have seen, was no doubt the origin of the mammalian plantigrade foot. Here, for instance, to all intents and purposes, is a similar type of foot, which belonged to a very early mammal, antecedent to the elephant series, the horse series, the rhinoceros, the hog, and, in short, all the known hoofed mammalia (Fig. 80). It was presumably an inhabitant of swampy ground, slow in its movements, and low in its intelligence.

But now, as we have seen, for more rapid progression on hard uneven ground, a stronger and better jointed foot would be needed. Therefore we find the bones of the wrist and ankle beginning to interlock, both among themselves and also with those of the foot and hand immediately below them. Such a stage of evolution is still apparent in the now existing elephant. (See Fig. 81.)

Next, however, a still stronger foot was made by the still further interlocking of the wrist and ankle bones, so that both the first and second rows of them were thus fitted into each other, as well as into the bones of the hand and foot beneath. This further modification is clearly traceable in some of the earlier perissodactyls, and occurs in the majority at the present time.

Compare, for example, the greater interlocking and consolidation of these small bones in the Rhinoceros as contrasted with the Elephant (Fig. 81). Moreover, simultaneously with these consolidating improvements in the mechanism of the wrist and ankle joints, or possibly at a somewhat later period, a reduction in the number of digits began to take place. This was a continuation of the policy of consolidating the foot, analogous to the dropping out of the sixth row of small bones in the paddle of Baptanodon. (Fig. 78.) In the pentadactyl plantigrade foot of the early mammals, the first digit, being the shortest, was the first to leave the ground, to dwindle, and finally to disappear. More work being thus thrown on the remaining four, they were strengthened by interlocking with the wrist (or ankle) bones above them, as just mentioned; and also by being brought closer together.

The changes which followed I will render in the words of Professor Marsh.

Two kinds of reduction began. One leading to the existing perissodactyl foot, and the other, apparently later, resulting in the artiodactyl type. In the former the axis of the foot remained in the middle of the third digit, as in the pentadactyl foot. [See Fig. 81.] In the latter, it shifted to the outer side of this digit, or between the third and fourth toe. [See Fig. 82.]

In the further reduction of the perissodactyl foot, the fifth digit, being shorter than the remaining three, next left the ground, and gradually disappeared. [Fig. 81 B.] Of the three remaining toes, the middle or axial one was the longest, and retaining its supremacy as greater strength and speed were required, finally assumed the chief support of the foot [Fig. 81 C], while the outer digits left the ground, ceased to be of use, and were lost, except as splint-bones [Fig. 81 D]. The feet of the existing horse shows the best example of this reduction in the Perissodactyls, as it is the most specialized known in the Ungulates [Fig. 81 D].

In the artiodactyl foot, the reduction resulted in the gradual diminution of the two outer of the four remaining toes, the third and fourth doing all the work, and thus increasing in size and power. The fifth digit, for the same reasons as in the perissodactyl foot, first left the ground and became smaller. Next, the second soon followed, and these two gradually ceased to be functional, [and eventually disappeared altogether, as shown in the accompanying drawing of the feet of still existing animals, Fig. 82 B, C, D].

The limb of the modern race-horse is a nearly perfect piece of machinery, especially adapted to great speed on dry, level ground. The limb of an antelope, or deer, is likewise well fitted for rapid motion on a plain, but the foot itself is adapted to rough mountain work as well, and it is to this advantage, in part, that the Artiodactyls owe their present supremacy. The plantigrade pentadactyl foot of the primitive Ungulate--and even the perissodactyl foot that succeeded it--both belong to the past humid period of the world's history. As the surface of the earth slowly dried up, in the gradual desiccation still in progress, new types of feet became a necessity, and the horse, antelope, and camel were gradually developed, to meet the altered conditions.

The best instance of such progressive modifications in the case of perissodactyl feet is furnished by the fossil pedigree of the existing horse, because here, within the limits of the same continuous family line, we have presented the entire series of modifications.

There are now known over thirty species of horse-like creatures, beginning from the size of a fox, then progressively increasing in bulk, and all standing in linear series in structure as in time. Confining attention to the teeth and feet, it will be seen from the wood-cut on page 189 that the former grow progressively longer in their sockets, and also more complex in the patterns of their crowns. On the other hand, the latter exhibit a gradual diminution of their lateral toes, together with a gradual strengthening of the middle one. (See Fig. 83.) So that in the particular case of the horse-ancestry we have a practically complete chain of what only a few years ago were "missing links." And this now practically completed chain shows us the entire history of what happens to be the most peculiar, or highly specialized, limb in the whole mammalian class--namely, that of the existing horse. Of the other two wood-cuts, the former (Fig. 84) shows the skeleton of a very early and highly generalized ancestor, while the other is a partial restoration of a much more recent and specialized one.

On the other hand, progressive modifications of the artiodactyl feet may be traced geologically up to the different stages presented by living ruminants, in some of which it has proceeded further than in others. For instance, if we compare the pig, the deer, and the camel (Fig. 82), we immediately perceive that the dwindling of the two rudimentary digits has proceeded much further in the case of the deer than in that of the pig, and yet not so far as in that of the camel, seeing that here they have wholly disappeared. Moreover, complementary differences are to be observed in the degree of consolidation presented by the two useful digits. For while in the pig the two foot-bones are still clearly distinguishable throughout their entire length, in the deer, and still more in the camel, their union is more complete, so that they go to constitute a single bone, whose double or compound character is indicated externally only by a slight bifurcation at the base. Nevertheless, if we examine the state of matters in the unborn young of these animals, we find that the two bones in question are still separated throughout their length, and thus precisely resemble what used to be their permanent condition in some of the now fossil species of hoofed mammalia.

Turning next from bones of the limb to other parts of the mammalian skeleton, let us briefly consider the evidence of evolution that is here likewise presented by the vertebral column, the skull, and the teeth.

As regards the vertebral column, if we examine this structure in any of the existing hoofed animals, we find that the bony processes called zygapophyses, which belong to each of the constituent vertebr? are so arranged that the anterior pair belonging to each vertebra interlocks with the posterior pair belonging to the next vertebra. In this way the whole series of vertebr?are connected together in the form of a chain, which, while admitting of considerable movement laterally, is everywhere guarded against dislocation. But if we examine the skeletons of any ungulates from the lower Eocene deposits, we find that in no case is there any such arrangement to secure interlocking. In all the hoofed mammals of this period the zygapophyses are flat. Now, from this flat condition to the present condition of full interlocking we obtain a complete series of connecting links. In the middle Miocene period we find a group of hoofed animals in which the articulation begins by a slight rounding of the previously flat surfaces: later on this rounding progressively increases, until eventually we get the complete interlocking of the present time.

As regards teeth, and still confining attention to the hoofed mammals, we find that low down in the geological series the teeth present on their grinding surfaces only three simple tubercles. Later on a fourth tubercle is added, and later still there is developed that complicated system of ridges and furrows which is characteristic of these teeth at the present time, and which was produced by manifold and various involutions of the three or four simple tubercles of Eocene and lower Miocene times. In other words, the principle of gradual improvement in the construction of teeth, which has already been depicted as regards the particular case of the Horse-family (Fig. 83), is no less apparent in the pedigree of all the other mammalia, wherever the paleontological history is sufficiently intact to serve as a record at all.

Lastly, as regards the skull, casts of the interior show that all the earlier mammals had small brains with comparatively smooth or unconvoluted surfaces; and that as time went on the mammalian brain gradually advanced in size and complexity. Indeed so small were the cerebral hemispheres of the primitive mammals that they did not overlap the cerebellum, while their smoothness must have been such as in this respect to have resembled the brain of a bird or reptile. This, of course, is just as it ought to be, if the brain, which the skull has to accommodate, has been gradually evolved into larger and larger proportions in respect of its cerebral hemispheres, or the upper masses of it which constitute the seat of intelligence. Thus, if we look at the above series of wood-cuts, which represents the comparative structure of the brain in the existing classes of the Vertebrata, we can immediately understand why the fossil skulls of Mammalia should present a gradual increase in size and furrowing, so as to accommodate the general increase of the brain in both these respects between the level marked "maml" and that marked "man," in the last of the diagrams.

The tabular statement on the following diagram, which I borrow from Prof. Cope, will serve at a glance to reveal the combined significance of so many lines of evidence, united within the limits of the same group of animals.

To give only one special illustration of the principle of evolution as regards the skull, here is one of the most recent instances that has occurred of the discovery of a missing link, or connecting form (see Fig. 88). The fossil (B), which was found in New Jersey, stands in an intermediate position between the stag and the elk. In the stag (A) the skull is high, showing but little of that anterior attenuation which is such a distinctive feature of the skull of the elk (C). The nasal bones (N) of the former, again, are remarkably long when compared with the similar bones of the latter, and the premaxillaries (PMX), instead of being projected forward along the horizontal plane of the base of the skull, are deflected sharply downward. In all these points, it will be seen, the newly discovered form (Cervalces) holds an intermediate position (B). "The skull exhibits a partial attenuation anteriorly, the premaxillaries are directed about equally downward and forward, and the nasal bones are measurably contracted in size. The horns likewise furnish characters which further serve to establish this dual relationship[18]."

[18] Heilprin, Geological Evidences of Evolution, pp. 73-4 (1888).

The evidence, then, which is furnished by all parts of the vertebral skeleton-- whether we have regard to Fishes, Reptiles, Birds, or Mammals--is cumulative and consistent. Nowhere do we meet with any deviation or ambiguity, while everywhere we encounter similar proofs of continuous transformation-- proofs which vary only with the varying amount of material which happens to be at our disposal, being most numerous and detailed in those cases where the greatest number of fossil forms has been preserved by the geological record. Here, therefore, we may leave the vertebral skeleton; and, having presented a sample of the evidence as yielded by horns and bones, I will conclude by glancing with similar brevity at the case of shells--which, as before remarked, constitute the only other sufficiently hard or permanent material to yield unbroken evidence touching the fossil ancestry of animals.

Of course it will be understood that I am everywhere giving merely samples of the now superabundant evidence which is yielded by Paleontology; and, as this chapter is already a long one, I must content myself with citing only the case of mollusk-shells, although shells of other classes might be made to yield highly important additions to the testimony. Moreover, even as regards the one division of mollusk-shells, I can afford to quote only a very few cases. These, however, are in my opinion the strongest single pieces of evidence in favour of transmutation which have thus far been brought to light.

Near the village of Steinheim, in Wetemberg, there is an ancient lake-basin, dating from Tertiary times. The lake has long ago dried up; but its aqueous deposits are extraordinarily rich in fossil shells, especially of different species of the genus Planorbis. The following is an authoritative summary of the facts.

As the deposits seem to have been continuous for ages, and the fossil shells very abundant, this seemed to be an excellent opportunity to test the theory of derivation. With this end in view, they have been made the subject of exhaustive study by Hilgendorf in 1866, and by Hyatt in 1880. In passing from the lowest to the highest strata the species change greatly and many times, the extreme forms being so different that, were it not for the intermediate forms, they would be called not only different species, but different genera. And yet the gradations are so insensible that the whole series is nothing less than a demonstration, in this case at least, of origin of species by derivation with modifications. The accompanying plate of successive forms (Fig. 89), which we take from Prof. Hyatt's admirable memoir, will show this better than any mere verbal explanation. It will be observed that, commencing with four slight varieties--probably sexually isolated varieties--of one species, each series shows a gradual transformation as we go upward in the strata--i. e. onward in time. Series I branches into three sub-series, in two of which the change of form is extreme. Series IV is remarkable for great increase in size as well as change in form. In the plate we give only selected stages, but in the fuller plates of the memoir, and still more in the shells themselves, the subtilest gradations are found[19].

[19] Le Conte, loc. cit., pp. 236-7.

Here is another and more recently observed case of transmutation in the case of mollusks.

The recent species, Strombus accipitrinus, still inhabits the coasts of Florida. Its extinct prototype, S. Leidy, was discovered a few years ago by Prof. Heilprin in the Pliocene formations of the interior of Florida. The peculiar shape of the wing, and tuberculation of the whorl, are thus proved to have grown but of a previously more conical form of shell.

Lastly, attention may here again be directed to the very instructive series of shells which has already been shown in a previous chapter, and which serves to illustrate the successive geological forms of Paludina from the Tertiary beds of Slavonia, as depicted by Prof. Neumayr of Vienna. (Fig. 1, p. 19.)

CHAPTER VI.

GEOGRAPHICAL DISTRIBUTION.

The argument from geology is the argument from the distribution of species in time. I will next take the argument from the distribution of species in space--that is, the present geographical distribution of plants and animals.

Seeing that the theory of descent with adaptive modification implies slow and gradual change of one species into another, and progressively still more slow and gradual changes of one genus, family, or order into another genus, family, or order, we should expect on this theory that the organic types living on any given geographical area would be found to resemble or to differ from organic types living elsewhere, according as the area is connected with or disconnected from other geographical areas. For instance, the large continental islands of Australia and New Zealand are widely disconnected from all other lands of the world, and deep sea soundings show that they have probably been thus disconnected, either since the time of their origin, or, at the least, through immense geological epochs. The theory of evolution, therefore, would expect to find two general facts with regard to the inhabitants of these islands. First, that the inhabitants should form, as it were, little worlds of their own, more or less unlike the inhabitants of any other parts of the globe. And next, that some of these inhabitants should present us with independent information touching archaic forms of life. For it is manifestly most improbable that the course of evolutionary history should have run exactly parallel in the case of these isolated oceanic continents and in continents elsewhere. Australia and New Zealand, therefore, ought to present a very large number, not only of peculiar species and genera, but even of families, and possibly of orders. Now this is just what Australia and New Zealand do present. The case of the dog being doubtful, there is an absence of all mammalian life, except that of one of the oldest and least highly developed orders, the Marsupials. There even occurs a unique order, still lower in the scale of organization--so low, in fact, that it deserves to be regarded as but nascent mammalian: I mean, of course, the Monotremata. As regards Birds, we have the peculiar wingless forms alluded to in a previous chapter (viz. that on Morphology); and, without waiting to go into details, it is notorious that the faunas of Australia and New Zealand are not only highly peculiar, but also suggestively archaic. Therefore, in both the respects above mentioned, the anticipations of our theory are fully borne out. But as it would take too long to consider, even cursorily, the faunas and floras of these immense islands, I here allude to them only for the sake of illustration. In order to present the argument from geographical distribution within reasonable limits, I think it is best to restrict our examination to smaller areas; for these will better admit of brief and yet adequate consideration. But of course it will be understood that the less isolated the region, and the shorter the time that it has been isolated, the smaller amount of peculiarity should we expect to meet with on the part of its present inhabitants. Or, conversely stated, the longer and the greater the isolation, the more peculiarity of species would our theory expect to find. The object of the present chapter will be to show that these, and other cognate expectations, are fully realized by facts; but, before proceeding to do this, I must say a few words on the antecedent standing of the argument.

Where the question is, as at present, between the rival theories of special creation and gradual transmutation, it may at first sight well appear that no test can be at once so crucial and so easily applied as this of comparing the species of one geographical area with those of another, in order to see whether there is any constant correlation between differences of type and degrees of separation. But a little further thought is enough to show that the test is not quite so simple or so absolute--that it is a test to be applied in a large and general way over the surface of the whole earth, rather than one to be relied upon as exclusively rigid in every special case.

In the first place, there is the obvious consideration that lands or seas which are discontinuous now may not always have been so, or not for long enough to admit of the effects of separation having been exerted to any considerable extent upon their inhabitants. Next, there is the scarcely less important consideration, that although land areas may long have been separated from one another by extensive tracts of ocean, birds and insects may more or less easily have been able to fly from one to the other; while even non-flying animals and plants may often have been transported by floating ice or timber, wind or water currents, and sundry other means of dispersal. Again, there is the important influence of climate to be taken into account. We know from geological evidence that in the course of geological time the self-same continents have been submitted to enormous changes of temperature-- varying in fact from polar cold to almost tropical heat; and as it is manifestly impossible that forms of life suited to one of these climates could have survived during the other, we can here perceive a further and most potent cause interfering with the test of geographical distribution as indiscriminately applied in all cases. When the elephant and hippopotamus were flourishing in England amid the luxuriant vegetation which these large animals require, it is evident that scarcely any one species of either the fauna or the flora of this country can have been the same as it was when its African climate gave place to that of Greenland. Therefore, as Mr. Wallace observes, "If glacial epochs in temperate lands and mild climates near the poles have, as now believed by men of eminence, occurred several times over in the past history of the earth, the effects of such great and repeated changes both on migration, modification, and extinction of species, must have been of overwhelming importance--of more importance perhaps than even the geological changes of sea and land."

But although for these, and certain other less important reasons which I need not wait to detail, we must conclude that the evidence from geographical distribution is not to be regarded as a crucial test between the rival theories of creation and evolution in all cases indiscriminately, I must next remark that it is undoubtedly one of the strongest lines of evidence which we possess. When we once remember that, according to the general theory of evolution itself, the present geographical distribution of plants and animals is "the visible outcome or residual product of the whole past history of the earth," and, therefore, that of the conditions determining the characters of life inhabiting this and that particular area continuity or discontinuity with other areas is but one,--when we remember this, we find that no further reservation has to be made: all the facts of geographical distribution speak with one consent in favour of the naturalistic theory.

* * * * *

The first of these facts which I shall adduce is, that although the geographical range of any given species is, as a rule, continuous, such is far from being always the case. Very many species have more or less discontinuous ranges--the mountain-hare, for instance, extending from the Arctic regions over the greater portion of Europe to the Ural Mountains and the Caucasus, and yet over all this enormous tract appearing only in isolated or discontinuous patches, where there happen to be either mountain ranges or climates cold enough to suit its nature. Now, in all such cases of discontinuity in the range of a species the theory of evolution has a simple explanation to offer--namely, either that some representatives of the species have at some former period been able to migrate from one region to the other, or else that at one time the species occupied the whole of the range in question, but afterwards became broken up as geographical, climatic, or other changes rendered parts of the area unfit for the species to inhabit. Thus, for instance, it is easy to understand that during the last cold epoch the mountain-hare would have had a continuous range; but that as the Arctic climate gradually receded to polar regions, the species would be able to survive in southern latitudes only on mountain ranges, and thus would become broken up into many discontinuous patches, corresponding with these ranges. In the same way we can explain the occurrence of Arctic vegetation on the Alps and Pyrenees--namely, as left behind by the retreat of the Arctic climate at the close of the glacial period.

But now, on the other hand, the theory of special creation cannot so well afford to render this obvious explanation of discontinuity. In the case of the Arctic flora of the Alps, for instance, although it is true that much of this vegetation is of an Arctic type, it is not true that the species are all identical with those which occur in the Arctic regions. Therefore the theory of special creation would here have to assume that, although the now common species were left behind on the Alps by the retreat of glaciation northwards, the peculiar Alpine species were afterwards created separately upon the Alps, and yet created with such close affinities to the pre-existing species as to be included with them under the same genera. Looking to the absurdity of this supposition, as well as of others which I need not wait to mention, certain advocates of special creation have sought to take refuge in another hypothesis--namely, that species which present a markedly discontinuous range may have had a corresponding number of different centres of creation, the same specific type having been turned down, so to speak, on widely separated areas. But to me it seems that this explanation presents even greater difficulty than the other. If it is difficult to say why the Divinity should have chosen to create new species of plants on the Alps on so precisely the same pattern as the old, much more would it be difficult to say why, in addition to these new species, he should also have created again the old species which he had already placed in the Arctic regions.

* * * * *

So much, then, for discontinuity of distribution. The next general fact to be adduced is, that there is no constant correlation between habitats and animals or plants suited to live upon them. Of course all the animals and plants living upon any given area are well suited to live upon that area; for otherwise they could not be there. But the point now is, that besides the area on which they do live, there are usually many other areas in different parts of the globe where they might have lived equally well--as is proved by the fact that when transported by man they thrive as well, or even better, than in their native country. Therefore, upon the supposition that all species were separately created in the countries where they are respectively found, we must conclude that they were created in only some of the places where they might equally well have lived. Probably there is at most but a small percentage either of plants or animals which would not thrive in some place, or places, on the earth's surface other than that in which they occur; and hence we must say that one of the objects of special creation--if this be the true theory--was that of depositing species in only some among the several parts of the earth's surface equally well suited to support them.

Now, I do not contend that this fact in itself raises any difficulty against the theory of special creation. But I do think that a very serious difficulty is raised when to this fact we add another--namely, that on every biological region we encounter species related to other species in genera, and usually also genera related to other genera in families. For if each of all the constituent species of a genus, and even of a family, were separately created, we must hence conclude that in depositing them there was an unaccountable design manifested to make areas of distribution correspond to the natural affinities of their inhabitants. For example, the humming-birds are geographically restricted to America, and number 120 genera, comprising over 400 species. Hence, if this betokens 400 separate acts of creation, it cannot possibly have been due to chance that they were all performed on the same continent: it must have been design which led to every species of this large family of birds having been deposited in one geographical area. Or, to take a case where only the species of a single genus are concerned. The rats and mice proper constitute a genus which comprises altogether more than 100 species, and they are all exclusively restricted to the Old World. In the New World they are represented by another genus comprising about 70 species, which resemble their Old World cousins in form and habits; but differ from them in dentition and other such minor points. Now, the question is,--Why should all the 100 species have been separately created on one side of the Atlantic with one pattern of dentition, and all the 70 species on the other side with another pattern? What has the Atlantic Ocean got to do with any "archetypal plan" of rats' teeth?

Or again, to recur to Australia, why should all the mammalian forms of life be restricted to the one group of Marsupials, when we know that not only the Rodents, such as the rabbit, but all other orders of mammals, would thrive there equally well. And similarly, of course, in countless other instances. Everywhere we meet with this same correlation between areas of distribution and affinities of classification.

Now, it is at once manifest how completely this general fact harmonizes with the theory of evolution. If the 400 species of humming-birds, for instance, are all modified descendants of common ancestors, and if none of their constituent individuals have ever been large enough to make their way across the oceans which practically isolate their territory from all other tropical and sub-tropical regions of the globe, then we can understand why it is that all the 400 species occupy the same continent. But on the special-creation theory we can see no reason why the 400 species should all have been deposited in America. And, as already observed, we must remember that this correlation between a geographically restricted habitat and the zoological or botanical affinities of its inhabitants, is repeated over and over and over again in the faunas and floras of the world, so that merely to enumerate the instances would require a separate chapter.

Furthermore, the general argument thus presented in favour of descent with continuous modification admits of being enormously strengthened by three different classes of additional facts.

The first is, that the correlation in question--namely, that between a geographically restricted habitat and the zoological or botanical affinities of its inhabitants--is not limited to the now existing species, but extends also to the extinct. That is to say, the dead species are allied to the living species, as we should expect that they must be, if the latter are modified descendants of the former. On the alternative theory, however, we have to suppose that the policy of maintaining a correlation between geographical restriction and natural affinity extends very much further back than even the existing species of plants and animals; indeed we must suppose that a practically infinite number of additional acts of separate creation were governed by the same policy, in the case of long lines of species long since extinct.

Thus far, then, the only answer which an advocate of special creation can adduce is, that for some reason unknown to us such a policy may have been more wise than it appears: it may have served some inscrutable purpose that allied products of distinct acts of creation should all be kept together on the same areas. Well, in answer to this unjustifiable appeal to the argument from ignorance, I will adduce the second of the three considerations. This is, that in cases where the geographical areas are not restricted the policy in question fails. In other words, where the inhabitants of an area are free to migrate to other areas, the policy of correlating affinity with distribution is most significantly forgotten. In this case species wander away from their native homes, and the course of their wanderings is marked by the origination of new species springing up en route. Now, is it reasonable to suppose that the mere circumstance of some members of a species being able to leave their native home should furnish any occasion for creating new and allied species upon the tracts over which they travel, or the territories to which they go? When the 400 existing species of humming-birds have all been created on the same continent for some reason supposed to be unknown, why should this reason give way before the accident of any means of migration being furnished to humming-birds, so that they should be able to visit, say the continents of Africa and Asia, there gain a footing beside the sun-birds, and henceforth determine a new centre for the separate creation of additional species of humming-birds peculiar to the Old World--as has happened in the case of the majority of species which, unlike the humming-birds, have been at any time free to migrate from their original homes?

Lastly, my third consideration is, that the supposed policy in question does not extend to affinities which are wider than those between species and genera--more rarely to families, scarcely ever to orders, and never to classes. In other words, nature shows a double correlation in her geographical distribution of organic types:--first, that which we have already considered between geographical restriction and natural affinity among inhabitants of the same areas; second, another of a more detailed character between degrees of geographical restriction and degrees of natural affinity. The more distant the affinity, the more general is the extension. This, of course, is what we should expect on the theory of descent with modification, because the more distant the affinity, and therefore, ex hypothesi, the larger and the older the original group of organisms, the greater must be the chance of dispersal. The 400 species of humming-birds may well be unable to migrate from their native continent; but it would indeed have been an unaccountable fact if no other species of all the class of birds had ever been able to have crossed the Atlantic Ocean. Thus, on the theory of evolution, we can well understand the second correlation now before us--namely, between remoteness of affinity and generality of dispersal,--so that there is no considerable portion of the habitable globe without representatives of all the classes of animals, few portions without representatives of all the orders, but many portions without many of the families, innumerable portions without innumerable genera, and, of course, all portions without the great majority of species. Now, while this general correlation thus obviously supports the theory of natural descent with progressive modification, it makes directly against the opposite theory of special creation. For we have recently seen that when we restrict our view to the case of species and genera, the theory of special creation is obliged to suppose that for some inscrutable reason the Deity had regard to systematic affinity while determining on what large areas to create his species[20]. But now we see that he must be held to have neglected this inscrutable reason (whatever it was) when he passed beyond the range of genera--and this always in proportion to the remoteness of systematic affinity on the part of the species concerned.

[20] I say "large areas" for the sake of argument; but the same correlation between distribution and affinity extends likewise to small areas where only small differences of affinity are concerned. Thus, for instance, speaking of smaller areas, Moritz Wagner says:--"The broader and more rapid the river, the higher and more regular the mountain-chain, the calmer and more extensive the sea, the more considerable, as a general rule, will be the taxonomic separation between the populations"; and he shows that, in correlation with such differences in the degrees of separation, are the degrees of diversification--i. e., the numbers of species, and even of varieties, which these topographical barriers determine.

I cannot well conceive a reductio ad absurdum more complete than this. But, having now presented these most general facts of geographical distribution in their relation to the issue before us, we may next proceed to consider a few illustrations of them in detail, for in this way I think that their overwhelming weight may become yet more abundantly apparent.

* * * * *

It will assist us in dealing with these detailed illustrations if we begin by considering the means of dispersal of organisms from one place to another. Of course the most ordinary means is that of continuous wandering, or emigration; but where geographical barriers of any kind have to be surmounted, organisms may only be able to pass them by more exceptional and accidental means. The principal barriers of a geographical kind are oceans, rivers, mountain-chains, and desert-tracts, in the case of terrestrial organisms; and, in the case of aquatic organisms, the presence of land. But it is to be observed that, as regards marine organisms, any considerable difference in the temperature of the water may constitute a barrier as effectual as the presence of land; and also that, in the case of all shallow-water faunas, a tract of deep ocean constitutes almost as complete a barrier as it does to terrestrial faunas.

Now, the means whereby barriers admit of being accidentally or occasionally surmounted are, of course, various; and they differ in the case of different organisms. Birds, bats, and insects, on account of their powers of flight, are particularly apt to be blown out great distances to sea, and hence of all animals are most likely to become the involuntary colonists of distant shores. Floating timber serves to convey seeds and eggs of small animals over great distances; and Darwin has shown that many kinds of seeds are able of themselves to float for more than a month in sea-water without losing their powers of germination. For instance, out of 87 kinds, 64 germinated after an immersion of 28 days, and a few survived an immersion of 137 days. As a result of all his experiments he concludes, that the seeds of at least ten per cent. of the species of plants of any country might be floated by sea-currents during 28 days, without losing their powers of germination; and this, at the average rate of flow of several Atlantic currents, would serve to transport the seeds to a distance of at least 900 miles. Again, he proved that even seeds which are quickly destroyed by contact with sea-water admit of being successfully transported during 30 days, if they be contained within the crop of a dead bird. He also proved that living birds are most active agents in the work of dissemination, and this not only by taking seeds into their crops (where, so long as they remain, the seeds are uninjured), but likewise by carrying seeds (and even young mollusks) attached to their feet and feathers. In the course of these experiments he found that a small cup-full of mud, which he gathered from the edges of three ponds in February, was so charged with seeds that when sown in the ground these few ounces of mud yielded no less than 537 plants, belonging to many different species. It is therefore evident what opportunities are thus afforded for the transportation of seeds on the feet and bills of wading-birds. Lastly, floating ice is well known to act as a carrier of any kind of life which may prove able to survive this mode of transit.

Such being the nature of geographical barriers, and the means that organisms of various kinds may occasionally have of overcoming them, I will now give a few detailed illustrations of the argument from geographical distribution, as previously presented in its general form.

To begin with aquatic animals. As Darwin remarks, "the marine inhabitants of the Eastern and Western shores of South America are very distinct; with extremely few shells, crustacea, or echinodermata in common." Again, westward of the shores of America, a wide space of open ocean extends, which, as we have seen, furnishes as effectual a barrier as does the land to any emigration of shallow-water animals. Now, as soon as this reach of deep water is passed, we meet in the eastern islands of the Pacific with another and totally distinct fauna. "So that three marine faunas range northward and southward in parallel lines not far from each other, under corresponding climates": they are, however, "separated from each other by impassable barriers, either of land or open sea": and it is in exact coincidence with the course of these barriers that we find so remarkable a differentiation of the faunas[21]. Obviously, therefore, it is impossible to suggest that this correlation is accidental. Altogether many thousands of species are involved, and within this comparatively limited area they are sharply marked off into three groups as to their natural affinities, and into three groups as to their several basins. Hence, if all these species were separately created, there is no escape from the conclusion that for some reason or another the act of creation was governed by the presence of these barriers, so that species deposited on the Eastern shores of South America were formed with one set of natural affinities, while species deposited on the Western shore were formed with another set; and similarly with regard to the third set of species in the third basin, which, extending over a whole hemisphere to the coast of

Africa without any further barrier, nowhere presents, over this vast area, any other case of a distinct marine fauna. But what conceivable reason can there have been thus to consult these geographical barriers in the original creation of specific types? Even if such a case stood alone, it would be strongly suggestive of error on the part of the special creation theory. But let us take another case, this time from fresh-water faunas.

[21] The only exception is in the case of the fish on each side of the isthmus of panama, where about 30 per cent, of the species are identical. But it is possible enough that at some previous time this narrow isthmus may have been even narrower than at present, if not actually open. At all events, the fact that this partial exception occurs just where the land-barrier is so narrow, is more suggestive of migration than of independent creation.

Although the geographical distribution of fresh-water fish and fresh-water shells is often surprisingly extensive and apparently capricious, this may be explained by the means of dispersal being here so varied--not only aquatic birds, floods, and whirlwinds, but also geographical changes of water-shed having all assisted in the process. Moreover, in some cases it is possible that the habits of more widely distributed fresh-water fish may have originally been wholly or partly marine--which, of course, would explain the existing discontinuity of their existing fresh-water distribution. But, be this as it may (and it is not a question that affects the issue between special creation and gradual evolution, since it is only a question as to how a given species has been dispersed from its original home, whether or not in that home it was specially created), the point I desire to bring forward is, that where we find a barrier to the emigration of fresh-water forms which is more formidable than a thousand miles of ocean--a barrier over which neither water-fowl nor whirlwinds are likely to pass, and which is above the reach of any geological changes of water-shed,--where we find such a barrier, we always find a marked difference in the fresh-water faunas on either side of it. The kind of barrier to which I allude is a high mountain-chain. It may be only a few miles wide; yet it exercises a greater influence on the diversification of specific types, where fresh-water faunas are concerned, than almost any other. But why should this be the case on any intelligible theory of special creation? Why, in the depositing of species of newly created fresh-water fish, should the presence of an impassable mountain-chain have determined so uniformly a difference of specific affinity on either side of it? The question, so far as I can see, does not admit of an answer from any reasonable opponent.

* * * * *

Turning now from aquatic organisms to terrestrial, the body of facts from which to draw is so large, that I think the space at my disposal may be best utilized by confining attention to a single division of them--that, namely, which is furnished by the zoological study of oceanic islands.

In the comparatively limited--but in itself extensive--class of facts thus presented, we have a particularly fair and cogent test as between the alternative theories of evolution and creation. For where we meet with a volcanic island, hundreds of miles from any other land, and rising abruptly from an ocean of enormous depth, we may be quite sure that such an island can never have formed part of a now submerged continent. In other words, we may be quite sure that it always has been what it now is--an oceanic peak, separated from all other land by hundreds of miles of sea, and therefore an area supplied by nature for the purpose, as it were, of testing the rival theories of creation and evolution. For, let us ask, upon these tiny insular specks of land what kind of life should we expect to find? To this question the theories of special creation and of gradual evolution would agree in giving the same answer up to a certain point. For both theories would agree in supposing that these islands would, at all events in large part, derive their inhabitants from accidental or occasional arrivals of wind-blown or water-floated organisms from other countries--especially, of course, from the countries least remote. But, after agreeing upon this point, the two theories must part company in their anticipations. The special-creation theory can have no reason to suppose that a small volcanic island in the midst of a great ocean should be chosen as the theatre of any extraordinary creative activity, or for any particularly rich manufacture of peculiar species to be found nowhere else in the world. On the other hand, the evolution theory would expect to find that such habitats are stocked with more or less peculiar species. For it would expect that when any organisms chanced to reach a wholly isolated refuge of this kind, their descendants should forthwith have started upon an independent course of evolutionary history. Protected from intercrossing with any members of their parent species elsewhere, and exposed to considerable changes in their conditions of life, it would indeed be fatal to the general theory of evolution if these descendants, during the course of many generations, were not to undergo appreciable change. It has happened on two or three occasions that European rats have been accidentally imported by ships upon some of these islands, and even already it is observed that their descendants have undergone a slight change of appearance, so as to constitute them what naturalists call local varieties. The change, of course, is but slight, because the time allowed for it has been so short. But the longer the time that a colony of a species is thus completely isolated under changed conditions of life the greater, according to the evolution theory, should we expect the change to become. Therefore, in all cases where we happen to know, from independent evidence of a geological kind, that an oceanic island is of very ancient formation, the evolution theory would expect to encounter a great wealth of peculiar species. On the other hand, as I have just observed, the special-creation theory can have no reason to suppose that there should be any correlation between the age of an oceanic island and the number of peculiar species which it may be found to contain.

Therefore, having considered the principles of geographical distribution from the widest or most general point of view, we shall pass to the opposite extreme, and consider exhaustively, or in the utmost possible detail, the facts of such distribution where the conditions are best suited to this purpose--that is, as I have already said, upon oceanic islands, which may be metaphorically regarded as having been formed by nature for the particular purpose of supplying naturalists with a crucial test between the theories of creation and evolution. The material upon which my analysis is to be based will be derived from the most recent works upon geographical distribution--especially from the magnificent contributions to this department of science which we owe to the labours of Mr. Wallace. Indeed, all that follows may be regarded as a condensed filtrate of the facts which he has collected. Even as thus restricted, however, our subject-matter would be too extensive to be dealt with on the present occasion, were we to attempt an exhaustive analysis of the floras and faunas of all oceanic islands upon the face of the globe. Therefore, what I propose to do is to select for such exhaustive analysis a few of what may be termed the most oceanic of oceanic islands--that is to say, those oceanic islands which are most widely separated from mainlands, and which, therefore, furnish the most unquestionable of test cases as between the theories of special creation and genetic descent.

* * * * *

Azores.--A group of volcanic islands, nine in number, about 900 miles from the coast of Portugal, and surrounded by ocean depths of 1,800 to 2,500 fathoms. There is geological evidence that the origin of the group dates back at least as far as Miocene times. There is a total absence of all terrestrial Vertebrata, other than those which are known to have been introduced by man. Flying animals, on the other hand, are abundant; namely, 53 species of birds, one species of bat, a few species of butterflies, moths, and hymenoptera, with 74 species of indigenous beetles. All these animals are unmodified European species, with the exception of one bird and many of the beetles. Of the 74 indigenous species of the latter, 36 are not found in Europe; but 19 are natives of Madeira or the Canaries, and 3 are American, doubtless transplanted by drift-wood. The remaining 14 species occur nowhere else in the world, though for the most part they are allied to other European species. There are 69 known species of land-shells, of which 37 are European, and 32 peculiar, though all allied to European forms. Lastly, there are 480 known species of plants, of which 40 are peculiar, though allied to European species.

Bermudas.--A small volcanic group of islands, 700 miles from North Carolina. Although there are about 100 islands in the group, their total area does not exceed 50 square miles. The group is surrounded by water varying in depth from 2,500 to 3,800 fathoms. The only terrestrial Vertebrate (unless the rats and mice are indigenous) is a lizard allied to an American form, but specifically distinct from it, and therefore a solitary species which does not occur anywhere else in the world. None of the birds or bats are peculiar, any more than in the case of the Azores; but, as in that case, a large percentage of the land-shells are so--namely, at least one quarter of the whole. Neither the botany nor the entomology of this group has been worked out; but I have said enough to show how remarkably parallel are the cases of these two volcanic groups of islands situated in different hemispheres, but at about the same distance from large continents. In both there is an extraordinary paucity of terrestrial vertebrata, and of any peculiar species of bird or beast. On the other hand, there is in both a marvellous wealth of peculiar species of insects and land-shells. Now these correlations are all abundantly intelligible. It is a difficult matter for any terrestrial animal to cross 900, or even 700, miles of ocean: therefore only one lizard has succeeded in doing so in one of the two parallel cases; and, living cut off from intercrossing with its parent form, the descendants of that lizard have become modified so as to constitute a peculiar species. But it is more easy for large flying animals to cross those distances of ocean: consequently, there is only one instance of a peculiar species of bird or bat--namely, a bull-finch in the Azores, which, being a small land-bird, is not likely ever to have had any other visitors from its original parent species coming over from Europe to keep up the original breed. Lastly, it is very much more easy for insects and land-mollusca to be conveyed to such islands by wind and floating timber than it is for terrestrial mammals, or even than it is for small birds and bats; but yet such means of transit are not sufficiently sure to admit of much recruiting from the mainland for the purpose of keeping up the specific types. Consequently, the insects and the land-shells present a much greater proportion of peculiar species--namely, one half and one fourth of the land-shells in the one case, and one eighth of the beetles in the other. All these correlations, I say, are abundantly intelligible on the theory of evolution; but who shall explain, on the opposite theory, why orders of beetles and land-mollusca should have been chosen from among all other animals for such superabundant creation on oceanic islands, so that in the Azores alone we find no less than 32 of the one and 14 of the other? And, in this connexion, I may again allude to the peculiar species of beetles in the island of Madeira. Here there are an enormous number of peculiar species, though they are nearly all related to, or included under the same genera as, beetles on the neighbouring continent. Now, as we have previously seen, no less than 200 of these species have lost the use of their wings. Evolutionists explain this remarkable fact by their general laws of degeneration under disuse, and the operation of natural selection, as will be shown later on; but it is not so easy for special creationists to explain why this enormous number of peculiar species of beetles should have been deposited on Madeira, all allied to beetles on the nearest continent, and nearly all deprived of the use of their wings. And similarly, of course, with all the peculiar species of the Bermudas and the Azores. For who will explain, on the theory of independent creation, why all the peculiar species, both of animals and plants, which occur on the Bermudas should so unmistakably present American affinities, while those which occur on the Azores no less unmistakably present European affinities? But to proceed to other, and still more remarkable, cases.

The Galapagos Islands.--This archipelago is of volcanic origin, situated under the equator between 500 and 600 miles from the West Coast of South

America. The depth of the ocean around them varies from 2,000 to 3,000 fathoms or more. This group is of particular interest, from the fact that it was the study of its fauna which first suggested to Darwin's mind the theory of evolution. I will, therefore, begin by quoting a short passage from his writings upon the zoological relations of this particular fauna.

Here almost every product of the land and of the water bears the unmistakeable stamp of the American continent. There are twenty-six land birds; of these, twenty-one, or perhaps twenty-three, are ranked as distinct species, and would commonly be assumed to have been here created; yet the close affinity of most of these birds to American species is manifest in every character, in their habits, gestures, and tones of voice. So it is with the other animals, and with a large proportion of the plants, as shown by Dr. Hooker in his admirable Flora of this archipelago. The naturalist, looking at the inhabitants of these volcanic islands in the Pacific, distant several hundred miles from the continent, feels that he is standing on American land. Why should this be so? Why should the species which are supposed to have been created in the Galapagos Archipelago, and nowhere else, bear so plainly the stamp of affinity to those created in America? There is nothing in the conditions of life, in the geological nature of the islands, in their height or climate, or in the proportions in which the several classes are associated together, which closely resembles the conditions of the South American coast; in fact, there is a considerable dissimilarity in all these respects. On the other hand, there is a considerable degree of resemblance in the volcanic nature of the soil, in the climate, height, and size of the islands, between the Galapagos and Cape de Verde Archipelagoes; but what an entire and absolute difference in their inhabitants! The inhabitants of the Cape de Verde Islands are related to those of Africa, like those of the Galapagos to America. Facts such as these admit of no sort of explanation on the ordinary view of independent creation; whereas on the view here maintained, it is obvious that the Galapagos Islands would be likely to receive colonists from America, and the Cape de Verde Islands from Africa; such colonists would be liable to modification--the principle of inheritance still betraying their original birthplace[22].

[22] origin of species, pp. 353-4.

The following is a synopsis of the fauna and flora of this archipelago, so far as at present known. The only terrestrial vertebrates are two peculiar species of land-tortoise, and one extinct species; five species of lizards, all peculiar--two of them so much so as to constitute a peculiar genus;--and two species of snakes, both closely allied to South American forms. Of birds there are 57 species, of which no less than 38 are peculiar; and all the non-peculiar species, except one, belong to aquatic tribes. The true land birds are represented by 31 species, of which all, except one, are peculiar; while more than half of them go to constitute peculiar genera. Moreover, while they are all unquestionably allied to South American forms, they present a beautiful series of gradations, "from perfect identity with the continental species, to genera so distinct that it is difficult to determine with what forms they are most nearly allied; and it is interesting to note that this diversity bears a distinct relation to the probabilities of, and facilities for, migration to the islands. The excessively abundant rice-bird, which breeds in Canada, and swarms over the whole United States, migrating to the West Indies and South America, visiting the distant Bermudas almost every year, and extending its range as far as Paraguay, is the only species of land-bird which remains completely unchanged in the Galapagos; and we may therefore conclude that some stragglers of the migrating host reach the islands sufficiently often to keep up the purity of the breed[23]." Again, of the thirty peculiar land-birds, it is observable that the more they differ from any other species or genera on the South American continent, the more certainly are they found to have their nearest relations among those South American forms which have the more restricted range, and are therefore the least likely to have found their way to the islands with any frequency.

[23] Wallace, Island Life, pp. 271-2.

The insect fauna of the Galapagos islands is scanty, and chiefly composed of beetles. These number 35 species, which are nearly all peculiar, and in some cases go to constitute peculiar genera. The same remarks apply to the twenty species of land-shells. Lastly, of the total number of flowering plants (332 species) more than one half (174 species) are peculiar. It is observable in the case of these peculiar species of plants--as also of the peculiar species of birds--that many of them are restricted to single islands. It is also observable that, with regard both to the fauna and flora, the Galapagos Islands as a whole are very much richer in peculiar species than either the Azores or Bermudas, notwithstanding that both the latter are considerably more remote from their nearest continents. This difference, which at first sight appears to make against the evolutionary interpretation, really tends to confirm it. For the Galapagos Islands are situated in a calm region of the globe, unvisited by those periodic storms and hurricanes which sweep over the North Atlantic, and which every year convey some straggling birds, insects, seeds, &c., to the Azores and Bermudas. Notwithstanding their somewhat greater isolation geographically, therefore, the Azores and Bermudas are really less isolated biologically than are the Galapagos Islands; and hence the less degree of peculiarity on the part of their endemic species. But, on the theory of special creation, it is impossible to understand why there should be any such correlation between the prevalence of gales and a comparative inertness of creative activity. And, as we have seen, it is equally impossible on this theory to understand why there should be a further correlation between the degree of peculiarity on the part of the isolated species, and the degree in which their nearest allies on the mainland are there confined to narrow ranges, and therefore less likely to keep up any biological communication with the islands.

St. Helena.--A small volcanic island, ten miles long by eight wide, situated in mid-ocean, 1100 miles from Africa, and 1800 from South America. It is very mountainous and rugged, bounded for the most part by precipices, rising from ocean depths of 17,000 feet, to a height above the sea-level of nearly 3,000. When first discovered it was richly clothed with forests; but these were all destroyed by human agency during the 16th, 17th and 18th centuries. The records of civilization present no more lamentable instance of this kind of destruction. From a merely pecuniary point of view the abolition of these primeval forests has proved an irreparable loss; but from a scientific point of view the loss is incalculable. These forests served to harbour countless forms of life, which extended at least from the Miocene age, and which, having found there an ocean refuge, survived as the last remnants of a remote geological epoch. In those days, as Mr. Wallace observes, St. Helena must have formed a kind of natural museum or vivarium of archaic species of all classes, the interest of which we can now only surmise from the few remnants of those remnants, which are still left among the more inaccessible portions of the mountain peaks and crater edges. These remnants of remnants are as follows.

There is a total absence of all indigenous mammals, reptiles, fresh-water fish, and true land-birds. There is, however, a species of plover, allied to one in

South Africa; but it is specifically distinct, and therefore peculiar to the island. The insect life, on the other hand, is abundant. Of beetles no less than 129 species are believed to be aboriginal, and, with one single exception, the whole number are peculiar to the island. "But in addition to this large amount of specific peculiarity (perhaps unequalled anywhere else in the world), the beetles of this island are remarkable for their generic isolation, and for the altogether exceptional proportion in which the great divisions of the order are represented. The species belong to 39 genera, of which no less than 25 are peculiar to the island; and many of these are such isolated forms that it is impossible to find their allies in any particular country[24]." More than two-thirds of all the species belong to the group of weevils--a circumstance which serves to explain the great wealth of beetle-population, the weevils being beetles which live in wood, and St. Helena having been originally a densely wooded island. This circumstance is also in accordance with the view that the peculiar insect fauna has been in large part evolved from ancestors which reached the island by means of floating timber; for, of course, no explanation can be suggested why special creation of this highly peculiar insect fauna should have run so disproportionately into the production of weevils. About two-thirds of the whole number of beetles, or over 80 species, show no close affinity with any existing insects, while the remaining third have some relations, though often very remote, with European and African forms. That this high degree of peculiarity is due to high antiquity is further indicated, according to our theory, by the large number of species which some of the types comprise. Thus, the 54 species of Cossonid?may be referred to three types; the 11 species of Bembidium form a group by themselves; and the Heteromera form two groups. "Now, each of these types may well be descended from a single species, which originally reached the island from some other land; and the great variety of generic and specific forms into which some of them have diverged is an indication, and to some extent a measure, of the remoteness of their origin[25]." But, on the counter-supposition that all these 128 peculiar species were separately created to occupy this particular island, it is surely unaccountable that they should thus present such an arborescence of natural affinities amongst themselves.

[24] Wallace, Island Life, p. 287.

[25] Wallace, Island Life, p. 287.

Passing over the rest of the insect fauna, which has not yet been sufficiently worked out, we next find that there are only 20 species of indigenous land-shells--which is not surprising when we remember by what enormous reaches of ocean the island is surrounded. Of these 20 species no less than 13 have become extinct, three are allied to European species, while the rest are so highly peculiar as to have no near allies in any other part of the globe. So that the land-shells tell exactly the same story as the insects.

Lastly, the plants likewise tell the same story. The truly indigenous flowering plants are about 50 in number, besides 26 ferns. Forty of the former and ten of the latter are peculiar to the island, and, as Sir Joseph Hooker tells us, "cannot be regarded as very close specific allies of any other plants at all" Seventeen of them belong to peculiar genera, and the others all differ so markedly as species from their congeners, that not one comes under the category of being an insular form of a continental species. So that with respect to its plants no less than with respect to its animals, we find that the island of St. Helena constitutes a little world of unique species, allied among themselves, but diverging so much from all other known forms that in many cases they constitute unique genera.

Sandwich Islands.--These are an extensive group of islands, larger than any we have hitherto considered--the largest of the group being about the size of Devonshire. The entire archipelago is volcanic, with mountains rising to a height of nearly 14,000 feet. The group is situated in the middle of the North Pacific, at a distance of considerably over 2,000 miles from any other land, and surrounded by enormous ocean depths. The only terrestrial vertebrata are two lizards, one of which constitutes a peculiar genus. There are 24 aquatic birds, five of which are peculiar; four birds of prey, two of which are peculiar; and 16 land-birds, all of which are peculiar. Moreover, these 16 land-birds constitute no less than 10 peculiar genera, and even one peculiar family of five genera. This is an amount of peculiarity far exceeding that of any other islands, and, of course, corresponds with the great isolation of this archipelago. The only other animals which have here been carefully studied are the land-shells, and these tell the same story as the birds. For there are no less than 400 species which are all, without any exception, peculiar; while about three-quarters of them go to constitute peculiar genera. Again, of the plants, 620 species are believed to be endemic; and of these 377 are peculiar, yielding no less than 39 peculiar genera.

* * * * *

Prejudice apart, I think we must all now agree that it is needless to continue further this line of proof. I have chosen the smallest and most isolated islands for the purposes of our present argument, first because these furnish the most crucial kind of test, and next because they best admit of being dealt with in a short space. But, if necessary, a vast amount of additional material could be furnished, not only from other small oceanic islands, but still more from the largest islands of the world, such as Australia and New Zealand. However, after the detailed inventories which have now been given in the case of some of the smaller islands most remote from mainlands, we may well be prepared to accept it as a general law, that wherever there is evidence of land-areas having been for a long time separated from other land-areas, there we meet with a more or less extraordinary profusion of unique species, often running up into unique genera. And, in point of fact, so far as naturalists have hitherto been able to ascertain, there is no exception to this general law in any region of the globe. Moreover, there is everywhere a constant correlation between the degree of this peculiarity on the part of the fauna and flora, and the time during which they have been isolated. Thus, for instance, among the islands which I have called into evidence, those that are at once the most isolated and give independent proofs of the highest antiquity, are the Galapagos Islands, the Sandwich Islands, and St. Helena. Now, if we apply the method of tabular analysis to these three cases, we obtain the following most astonishing results. For the sake of simplicity I will omit the enumeration of peculiar genera, and confine attention to peculiar species. Moreover, I will consider only terrestrial animals; for, as we have already seen, aquatic animals are so much more likely to reach oceanic islands that they do not furnish nearly so fair a test of the evolutionary hypothesis.

PECULIAR SPECIES.

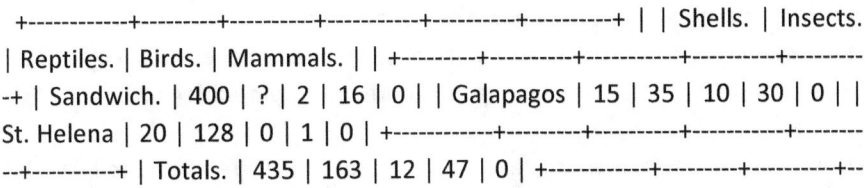

	Shells.	Insects.	Reptiles.	Birds.	Mammals.
Sandwich.	400	?	2	16	0
Galapagos	15	35	10	30	0
St. Helena	20	128	0	1	0
Totals.	435	163	12	47	0

NON-PECULIAR SPECIES.

	Reptiles.	Birds.	Mammals.	Shells.	Insects.
Sandwich.	0	?	0	0	0
Galapagos	?	?	0	1	0
St. Helena	0	?	0	0	0
Totals.	0	?	0	1	0

From this synopsis we perceive that out of a total of 658 species of terrestrial animals known to inhabit these three oceanic territories, all are peculiar, with the exception of a single land-bird which is found in the Galapagos Islands. This is the rice-bird, so very abundant on the American continent that its representatives must not unfrequently become the involuntary colonists of the Archipelago. There are, however, a few species of non-peculiar insects inhabiting the Sandwich and Galapagos Islands, the exact number of which is doubtful, and on this account are not here quoted. But at most they would be represented by units, and therefore do not affect the general result. Lastly, the remarkable fact will be noted, that there is no single representative of the mammalian class in any of these islands.

If we turn next to consider the case of plants, we obtain the following result:--

	Peculiar Species.	Non-peculiar Species.
Sandwich	377	243
Galapagos	174	158
St. Helena	50	26
Totals	601	427

So that by adding together peculiar species both of land-animals and plants, we find that on these three limited areas alone there are 1258 forms of life which occur nowhere else upon the globe--not to speak of the peculiar aquatic species, nor of the presumably large number of peculiar species of all kinds not hitherto discovered in these imperfectly explored regions.

Now let us compare these facts with those which are presented by the faunas and floras of islands less remote from continents, and known from independent geological evidence to be of comparatively recent origin--that is, to have been separated from their adjacent mainlands in comparatively recent times, and therefore as islands to be comparatively young. The British Isles furnish as good an instance as could be chosen, for they together comprise over 1000 islands of various sizes, which are nowhere separated from one another by deep seas, and in the opinion of geologists were all continuous with the European continent since the glacial period.

BRITISH ISLES.

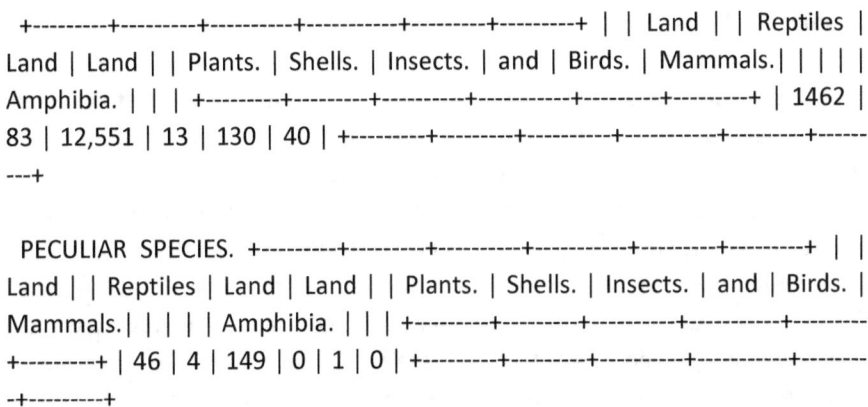

NON-PECULIAR SPECIES.

Plants.	Land Shells.	Insects.	Reptiles and Amphibia.	Birds.	Land Mammals.
1462	83	12,551	13	130	40

PECULIAR SPECIES.

Plants.	Land Shells.	Insects.	Reptiles and Amphibia.	Birds.	Land Mammals.
46	4	149	0	1	0

Total Peculiar Plants 46 Total Peculiar Animals 154 ---- Grand Total 200

I have drawn up this table in the most liberal manner possible, including as peculiar species forms which many naturalists regard as merely local varieties. But, even as thus interpreted, how wonderful is the contrast between the 1000 islands of Great Britain and the single volcanic rock of St. Helena, where almost all the animals and about half the plants are peculiar, instead of about 1/80 of the animals, and 1/30 of the plants. Of course, if no peculiar species of any kind had occurred in the British Isles, advocates of special creation might have argued that it was, so to speak, needless for the Divinity to have added any new species to those European forms which fully populated the islands at the time when they were separated from the continent. But, as the matter stands, advocates of special creation must face the fact that a certain small number of new and peculiar species have been formed on the British Isles; and, therefore, that creative activity has not been wholly suspended in their case. Why, then, has it been so meagre in this case of a thousand islands, when it has proved so profuse in the case of all single islands more remote from mainlands, and presenting a higher antiquity? Or why should the Divinity have thus appeared so uniformly to consult these merely accidental circumstances of space and time in the depositing of his unique specific types? Do not such facts rather speak with irresistible force in favour of the view, that while all ancient and solitary islands have had time enough, and separation enough, to admit of distinct histories of evolution having been written in their living inhabitants, no one of the thousand islands of Great Britain has had either time enough, or separation enough, to have admitted of more than some of the first pages of such a history having been commenced?

But this allusion to Great Britain introduces us to another point. It will have been observed that, unlike oceanic islands remote from mainlands, Great Britain is well furnished both with reptiles (including amphibia) and mammals. For there is no instance of any oceanic island situated at more than 300 miles from a continent where any single species of the whole class of mammals is to be found, excepting species of the only order which is able to fly--namely, the bats. And the same has to be said of frogs, toads, and newts, whose spawn is quickly killed by contact with sea-water, and therefore could never have reached remote islands in a living state. Hence, on evolutionary principles; it is quite intelligible why oceanic islands should not present any species of mammals or batrachians--peculiar or otherwise,--save such species of mammals as are able to fly. But on the theory of special creation we can assign no reason why, notwithstanding the extraordinary profusion of unique types of other kinds which we have seen to occur on oceanic islands, the Deity should have made this curious exception to the detriment of all frogs, toads, newts, and mammals, save only such as are able to fly. Or, if any one should go so far to save a desperate hypothesis as to maintain that there must have been some hidden reason why batrachians and quadrupeds were not specially created on oceanic islands, I may mention another small--but in this relation a most significant--fact. This is that on some of these islands there occur certain peculiar species of plants, the seeds of which are provided with numerous tiny hooks, obviously and beautifully adapted--like those on the seeds of allied plants elsewhere--to catch the wool or hair of moving quadrupeds, and so to further their own dissemination. But, as we have just seen, there are no quadrupeds in the islands to meet these beautiful adaptations on the part of the plants; so that special creationists must resort to the almost impious supposition that in these cases the Deity has only carried out half his plan, in that while he made an elaborate provision for these uniquely created species of plants, which depended for its efficiency on the presence of quadrupeds, he nevertheless neglected to place any quadrupeds on the islands where he had placed the plants. Such one-sided attempts at adaptation surely resolve the thesis of special creation to a reductio ad absurdum; and hence the only reasonable interpretation of them is, that while the seeds of allied or ancestral plants were able to float to the islands, no quadrupeds were ever able over so great a distance to swim.

* * * * *

Although much more evidence might still be given under the head of geographical distribution, I must now close with a brief summary of the main points that have been adduced.

After certain preliminary considerations, I began by noticing that the theory of evolution has a much more intelligible account to give than has its rival of the facts of discontinuous distribution--the Alpine flora, for instance, being allied to the Arctic, not because the same species were separately created in both places, but because during the glacial period these species extended all over Europe, and were left behind on the Alps as the Arctic flora receded northwards--which was sufficiently long ago to explain why some of the Alpine species are unique, though closely allied to Arctic forms.

Next we saw that, although living things are always adapted to the climates under which they live (since otherwise they could not live there at all), it is equally true that, as a rule, besides the area on which they do live, there are many other areas in different parts of the globe where they might have lived equally well. Consequently we must conclude that, if all species were separately created, many species were severally created on only one among a number of areas where they might equally well have thrived. Now, although this conclusion in itself may not seem opposed to the theory of special creation, a most serious difficulty is raised when it is taken in connexion with another fact of an equally general kind. This is, that on every biological region we encounter chains of allied species constituting allied genera, families, and so on; while we scarcely ever meet with allied species in different biological regions, notwithstanding that their climates may be similar, and, consequently, just as well suited to maintain some of the allied species. Hence we must further conclude, if all species were separately created, that in the work of creation some unaccountable regard was paid to making areas of distribution correspond to degrees of structural affinity. A great many species of the rat genus were created in the Old World, and a great many species of another, though allied, genus were created in the New World: yet no reason can be assigned why no one species of the Old World series should not just as well have been deposited in the New World, and vice versa. On the other hand, the theory of evolution may claim as direct evidence in its support all the innumerable cases such as these--cases, indeed, so innumerable that, as Mr. Wallace remarks, it may be taken as a law of nature that "every species has come into existence coincident both in space and time with a pre-existing and closely allied species." A general law which, while in itself most strongly suggestive of evolution, is surely impossible to reconcile with any reasonable theory of special creation. Furthermore, this law extends backwards through all geological time, with the result that the extinct species which now occur only as fossils on any given geological area, resemble the species still living upon that area, as we should expect that they must, if the former were the natural progenitors of the latter. On the other hand, if they were not the natural progenitors, but all the species, both living and extinct, were the supernatural and therefore independent creations which the rival theory would suppose, then no reason can be given why the extinct species should thus resemble the living--any more than why the living species should resemble one another. For, as we have seen, there are almost always many other habitats on other parts of the globe, where any members of any given group of species might equally well have been deposited; and this, of course, applies to geological no less than to historical time. Yet throughout all time we meet with this most suggestive correlation between continuity of a geographical area and structural affinity between the forms of life which have lived, or are still living, upon that area.

Similarly, we find the further, and no less suggestive, correlation between the birth of new species and the immediate pre-existence of closely allied species on the same area--or, at most, on closely contiguous areas.

Where a continuous area has long been circumscribed by barriers of any kind, which prevent the animals from wandering beyond it, then we find that all the species, both extinct and living, constitute more or less a world of their own; while, on the other hand, where the animals are free to migrate from one area to another, the course of their migrations is marked by the origination of new species springing up en route, and serving to connect the older, or metropolitan, forms with the younger, or colonising, forms in the way of a graduated series. This principle, however, admits of being traced only in certain cases of species belonging to the same genus, of genera belonging to the same family, or, at most, of families belonging to the same order. In other words, the more general the structural affinity, the more general is the geographical extension--as we should expect to be the case on the theory of descent with branching modifications, seeing that the larger, the older, and the more diverse the group of organisms compared, the greater must be their chances of dispersal.

These general considerations led us to contemplate more in detail the correlation between structural affinity and barriers to free migration. Such barriers, of course, differ in the cases of different organisms. Marine organisms are stopped by land, unsuitable temperature, or unsuitable depths; fresh-water organisms by sea and by mountain-chains; terrestrial organisms chiefly by water. Now it is a matter of fact which admits of no dispute, that in each of these cases we meet with a direct correlation between the kind of barrier and the kind of organisms whose structural affinities are affected thereby. Where we have to do with marine organisms, barriers such as the Isthmus of Panama and the varying depth of the Western Pacific determine three very distinct faunas, ranging north and south in closely parallel lines, and under corresponding climates. Where we have to do with fresh-water organisms, we find that a mountain-chain only a few miles wide has more influence in determining differences of organic type on either side of it than is exercised by even thousands of miles of a continuous land-area, if this be uninterrupted by any mountains high enough to prevent water-fowl, whirlwinds, &c., from dispersing the ova. Again, where we have to do with terrestrial organisms, the most effectual barriers are wide reaches of ocean; and, accordingly, we find that these exercise an enormous influence on the modification of terrestrial types. Moreover, we find that the more terrestrial an organism, or the greater the difficulty it has in traversing a wide reach of ocean, the greater is the modifying influence of such a barrier upon that type.

In oceanic islands, for example, many of the plants and aquatic birds usually belong to the same species as those which occur on the nearest mainlands, and where there are any specific differences, these but rarely run up to generic differences. But the land-birds, insects, and reptiles which are found on such islands are nearly always specifically, and very often generically, distinct from those on the nearest mainland--although invariably allied with sufficient closeness to leave no manner of doubt as to their affinities with the fauna of that mainland. Lastly, no amphibians and no mammals (except bats) are ever found on any oceanic islands. Yet, as we have seen, on the theory of special creation, these islands must all be taken to have been the theatres of the most extraordinary creative activity, so that on only three of them we found no less than 1258 unique species, whereof 657 were unique species of land animals, to be set against one single species known to occur elsewhere. Nevertheless, notwithstanding this prodigious expenditure of creative energy in the case of land-birds, land-shells, insects, and reptiles, no single new amphibian, or no single new mammal, has been created on any single oceanic island, if we except the only kind of mammal that is able to fly, and the ancestors of which, like those of the land-birds and insects, might therefore have reached the islands ages ago. Moreover, with regard to mammals, even in cases where allied forms occur on either side of a sea-channel, it is found to be a general rule that if the channel is shallow, the species on either side of it are much more closely related than if it be deep--and this irrespective of its width. Therefore we can only conclude, in the words of Darwin--"As the amount of modification which animals of all kinds undergo partly depends on lapse of time, and as the islands which are separated from each other or from the mainland by shallow channels are more likely to have been continuously united within a recent period than islands separated by deeper channels, we can understand how it is that a relation exists between the depth of the sea separating two mammalian faunas, and the degree of their affinity--a relation which is quite inexplicable on the theory of independent acts of creation."

* * * * *

Looking to all these general principles of geographical distribution, and remembering the sundry points of smaller detail relating to oceanic islands which I will not wait to recapitulate, to my mind it seems that there is no escape from the following conclusion, with which I will bring my brief epitome of the evidence to a close. The conclusion to which, I submit, all the evidence leads is, that if the doctrine of special creation is taken to be true, then it must be further taken that the one and only principle which has been consistently followed in the geographical deposition of species, is that of so depositing them as to make it everywhere appear that they were not thus deposited at all, but came into existence where they now occur by way of genetic descent with perpetual migration and correlative modification. On no other principle, so far as I can see, would it be possible to account for the fact that "every species has come into existence coincident both in space and time with a pre-existing and closely allied species," together with the carefully graduated regard to physical barriers which the Creator must have displayed while depositing his newly formed species on either sides of them-- everywhere making degrees of structural affinity correspond to degrees of geographical continuity, and degrees of structural difference correspond to degrees of geographical separation, whether by mountain-chains in the case of fresh-water faunas, by land and by deep sea in the case of marine faunas, or by reaches of ocean in the case of terrestrial faunas--stocking oceanic islands with an enormous profusion of peculiar species all allied to those on the nearest mainlands, yet everywhere avoiding the creation upon them of any amphibian or mammal, except an occasional bat. We are familiar with the doctrine that God is a God who hideth himself; here, however, it seems to me, we should have but a thinly-veiled insinuation, not merely that in his works he is hidden, but that in these works he is untrue. Than which I cannot conceive a stronger condemnation of the theory which it has been my object fairly to represent and dispassionately to criticise.

SECTION II

SELECTION

CHAPTER VII.

THE THEORY OF NATURAL SELECTION.

Thus far we have been considering the main evidences of organic evolution considered as a fact. We now enter a new field, namely, the evidences which thus far have been brought to light touching the causes of organic evolution considered as a process.

As was pointed out in the opening chapter, this is obviously the methodical course to follow: we must have some reasonable assurance that a fact is a fact before we endeavour to explain it. Nevertheless, it is not necessary that we should actually demonstrate a fact to be a fact before we endeavour to explain it. Even if we have but a reasonable presumption as to its probability, we may find it well worth while to consider its explanation; for by so doing we may obtain additional evidence of the fact itself. And this because, if it really is a fact, and if we hit upon the right explanation of it, by proving the explanation probable, we may thereby greatly increase our evidence of the fact. In the very case before us, for example, the evidence of evolution as a fact has from the first been largely derived from testing Darwin's theory concerning its method. It was this theoretical explanation of its method which first set him seriously to enquire into the evidences of evolution as a fact; and ever since he published his results, the evidences which he adduced in favour of natural selection as a method have constituted some of the strongest reasons which scientific men have felt for accepting evolution as a fact. Of course the evidence in favour of this fact has gone on steadily growing, quite independently of the assistance which was thus so largely lent to it by the distinctively Darwinian theory of its method; and, indeed, so much has this been the case, that in the present treatise we have been able to consider such direct evidence of the fact itself, without any reference at all to the indirect or accessory evidence which is derived from that of natural selection as a method. From which it follows that in most of what I am about to say in subsequent chapters on the evidences of natural selection as a method, there will be furnished a large addition to the evidences which have already been detailed of evolution as a fact. But, as a matter of systematic treatment, I have thought it desirable to keep these two branches of our subject separate. Which means that I have made the evidences of evolution as a fact to stand independently on their own feet--feet which in my opinion are amply strong enough to bear any weight of adverse criticism that can be placed upon them.

Our position, then, is this. On the foundation of the previous chapters, I will henceforth assume that we all accept organic evolution as a fact, without requiring any of the accessory evidence which is gained by independent proof of natural selection as a method. But in making this assumption--namely, that we are all now firmly persuaded of the fact of evolution--I do not imagine that such is really the case. I make the assumption for the purposes of systematic exposition, and in order that different parts of the subject may be kept distinct. I confess it does appear to me remarkable that there should still be a doubt in any educated mind touching the general fact of evolution; while it becomes to me unaccountable that such should be the case with a few still living men of science, who cannot be accused of being ignorant of the evidences which have now been accumulated. But in whatever measure we may severally have been convinced--or remained unconvinced--on this matter, for the purposes of exposition I must hereafter assume that we are all agreed to the extent of regarding the process of evolution as, at least, sufficiently probable to justify enquiry touching its causes on supposition of its truth.

Now, the causes of evolution have been set forth in a variety of different hypotheses, only the chief of which need be mentioned here. Historically speaking the first of these was that which was put forward by Erasmus Darwin, Lamarck, and Herbert Spencer. It consists in putting together the following facts and inferences.

We know that, in the lifetime of the individual, increased use of structures leads to an increase of their functional efficiency; while, on the other hand, disuse leads to atrophy. The arms of a blacksmith, and the legs of a mountaineer, are familiar illustrations of the first principle: our hospital wards are full of illustrations of the second. Again, we know that the characters of parents are transmitted to their progeny by means of heredity. Now the hypothesis in question consists in supposing that if any particular organs in a species are habitually used for performing any particular action, they must undergo a structural improvement which would more and more adapt them to the performance of that action; for in each generation constant use would better and better adapt the structures to the discharge of their functions, and they would then be bequeathed to the next generation in this their improved form by heredity. So that, for instance, if there had been a thousand generations of blacksmiths, we might expect the sons of the last of them to inherit unusually strong arms, even if these young men had themselves taken to some other trade not requiring any special use of their arms. Similarly, if there had been a thousand generations of men who used their arms but slightly, we should expect their descendants to show but a puny development of the upper extremities. Now let us apply all this to the animal kingdom in general. The giraffe, for instance, is a ruminant whose entire frame has been adapted to support an enormously long neck, which is of use to the animal in reaching the foliage of trees. The ancestors of the giraffe, having had ordinary necks, were supposed by Lamarck to have gradually increased the length of them, through many successive generations, by constantly stretching to reach high foliage; and he further supposed that, when the neck became so long as to require for its support special changes in the general form of the animal as a whole, these special changes would have brought about the dwindling of other parts from which so much activity was no longer required--the general result being that the whole organization of the animal became more and more adapted to browsing on high foliage. And so in the cases of other animals, Lamarck believed that the adaptation of their forms to their habits could be explained by this simple hypothesis that the habits created the forms, through the effects of use and disuse, coupled with heredity.

Such is what is ordinarily known as Lamarck's theory of evolution. We may as well remember, however, that it really constitutes only one part of his theory; for besides this hypothesis of the cumulative inheritance of functionally-produced modifications--to which we may add the inherited effects of any direct action exercised by surrounding conditions of life,-- Lamarck believed in some transcendental principle tending to produce gradual improvement in pre-determined lines of advance. Therefore it would really be more correct to designate the former hypothesis by the name either of Erasmus Darwin, or, still better, of Herbert Spencer. Nevertheless, in order to avoid confusion, I will follow established custom, and subsequently speak of this hypothesis as the Lamarckian hypothesis--understanding, however, that in employing this designation I am not referring to any part or factor of Lamarck's general theory of evolution other than the one which has just been described--namely, the hypothesis of the cumulative transmission of functionally-produced, or otherwise "acquired," modifications.

This, then, was the earliest hypothesis touching the causes of organic evolution. But we may at once perceive that it is insufficient to explain all that stands to be explained. In the first place, it refers in chief part only to the higher animals, which are actuated to effort by intelligence. Its explanatory power in the case of most invertebrata--as well as in that of all plants--is extremely limited, inasmuch as these organisms can never be moved to a greater or less use of their several parts by any discriminating volition, such as that which leads to the continued straining of a giraffe's neck for the purpose of reaching foliage. In the second place, even among the higher animals there are numberless tissues and organs which unquestionably present a high degree of adaptive evolution, but which nevertheless cannot be supposed to have fallen within the influence of Lamarckian principles. Of such are the shells of crustacea, tortoises, &c., which although undoubtedly of great use to the animals presenting them, cannot ever have been used in the sense required by Lamarck's hypothesis, i. e. actively exercised, so as to increase a flow of nutrition to the part. Lastly, in the third place, the validity of Lamarck's hypothesis in any case whatsoever has of late years become a matter of serious question, as will be fully shown and discussed in the next volume. Meanwhile it is enough to observe that, on account of all these reasons, the theory of Lamarck, even if it be supposed to present any truth at all, is clearly insufficient as a full or complete theory of organic evolution.

* * * * *

In historical order the next theory that was arrived at was the theory of natural selection, simultaneously published by Darwin and Wallace on July 1st, 1858.

If we may estimate the importance of an idea by the change of thought which it effects, this idea of natural selection is unquestionably the most important idea that has ever been conceived by the mind of man. Yet the wonder is that it should not have been hit upon long before. Or rather, I should say, the wonder is that its immense and immeasurable importance should not have been previously recognised. For, since the publication of this idea by Darwin and Wallace, it has been found that its main features had already occurred to at least two other minds--namely, Dr. Wells in 1813, and Mr. Patrick Matthew in 1831. But neither of these writers perceived that in the few scattered sentences which they had written upon the subject they had struck the key-note of organic nature, and resolved one of the principal chords of the universe. Still more remarkable is the fact that Mr. Herbert Spencer--notwithstanding his great powers of abstract thought and his great devotion of those powers to the theory of evolution, when as yet this theory was scorned by science--still more remarkable, I say, is the fact that Mr. Herbert Spencer should have missed what now appears so obvious an idea. But most remarkable of all is the fact that Dr. Whewell, with all his stores of information on the history of the inductive sciences, and with all his acumen on the matter of scientific method, should not only have conceived the idea of natural selection, but expressly stated it as a logically possible explanation of the origin of species, and yet have so stated it merely for the purpose of dismissing it with contempt[26]. This, I think, is most remarkable, because it serves to prove how very far men's minds at that time must have been from entertaining, as in any way antecedently probable, the doctrine of transmutation. In order to show this I will here quote one passage from the writings of Whewell, and another from a distinguished French naturalist referred to by him.

[26] For quotations, see Note A.

In 1846 Whewell wrote:--

Not only is the doctrine of the transmutation of species in itself disproved by the best physiological reasonings, but the additional assumptions which are requisite to enable its advocates to apply it to the explanation of the geological and other phenomena of the earth, are altogether gratuitous and fantastical[27].

[27] whewell, indications of the creator, 2nd ed., 1846.

Then he quotes with approval the following opinion:--

Against this hypothesis, which, up to the present time, I regard as purely gratuitous, and likely to turn geologists out of the sound and excellent road in which they now are, I willingly raise my voice, with the most absolute conviction of being in the right[28].

[28] de blainville, compte rendu, 1837.

And, after displaying the proof rendered by Lyell of uniformitarianism in geology, and cordially subscribing thereto, Whewell adds:--

We are led by our reasonings to this view, that the present order of things was commenced by an act of creative power entirely different to any agency which has been exerted since. None of the influences which have modified the present races of animals and plants since they were placed in their habitations on the earth's surface can have had any efficacy in producing them at first. We are necessarily driven to assume, as the beginning of the present cycle of organic nature, an event not included in the course of nature[29].

[29] Whewell, ibid., p. 162.

So much, then, for the state of the most enlightened and representative opinions on the question of evolution before the publication of Darwin's work; and so much, likewise, for the only reasonable suggestions as to the causes of evolution which up to that time had been put forward, even by those few individuals who entertained any belief in evolution as a fact. It was the theory of natural selection that changed all this, and created a revolution in the thought of our time, the magnitude of which in many of its far-reaching consequences we are not even yet in a position to appreciate; but the action of which has already wrought a transformation in general philosophy, as well as in the more special science of biology, that is without a parallel in the history of mankind.

* * * * *

Although every one is now more or less well acquainted with the theory of natural selection, it is necessary, for the sake of completeness, that I should state the theory; and I will do so in full detail.

It is a matter of observable fact that all plants and animals are perpetually engaged in what Darwin calls a "struggle for existence." That is to say, in every generation of every species a great many more individuals are born than can possibly survive; so that there is in consequence a perpetual battle for life going on among all the constituent individuals of any given generation. Now, in this struggle for existence, which individuals will be victorious and live? Assuredly those which are best fitted to live, in whatever respect, or respects, their superiority of fitness may consist. Hence it follows that Nature, so to speak, selects the best individuals out of each generation to live. And not only so; but as these favoured individuals transmit their favourable qualities to their offspring, according to the fixed laws of heredity, it further follows that the individuals composing each successive generation have a general tendency to be better suited to their surroundings than were their forefathers. And this follows, not merely because in every generation it is only the "flower of the flock" that is allowed to breed, but also because, if in any generation some new and beneficial qualities happen to arise as slight variations from the ancestral type, they will (other things permitting) be seized upon by natural selection, and, being transmitted by heredity to subsequent generations, will be added to the previously existing type. Thus the best idea of the whole process will be gained by comparing it with the closely analogous process whereby gardeners, fanciers, and cattle-breeders create their wonderful productions; for just as these men, by always "selecting" their best individuals to breed from, slowly but continuously improve their stock, so Nature, by a similar process of "selection" slowly but continuously makes the various species of plants and animals better and better suited to the conditions of their life.

Now, if this process of continuously adapting organisms to their environment takes place in nature at all, there is no reason why we should set any limits on the extent to which it is able to go, up to the point at which a complete and perfect adaptation is achieved. Therefore we might suppose that all species would eventually reach this condition of perfect harmony with their environment, and then remain fixed. And so, according to the theory, they would, if the environment were itself unchanging. But forasmuch as the environment (i. e. the sum total of the external conditions of life) of almost every organic type alters more or less from century to century--whether from astronomical, geological, and geographical changes, or from the immigrations and emigrations of other species living on contiguous areas, and so on--it follows that the process of natural selection need never reach a terminal phase. And forasmuch as natural selection may thus continue, ad infinitum, slowly to alter a specific type in adaptation to a gradually changing environment, if in any case the alteration thus effected is sufficient in amount to lead naturalists to name the result as a distinct species, it follows that natural selection has transmuted one specific type into another. Similarly, by a continuation of the process, specific types would become transmuted into generic, generic into family types, and so on. Thus the process is supposed to go on throughout all the countless forms of life continuously and simultaneously--the world of organic types being thus regarded as in a state of perpetual, though gradual, flux.

* * * * *

Now, the first thing we have to notice about this theory is, that in all its main elements it is merely a statement of observable facts. It is an observable fact that in all species of plants and animals a very much larger number of individuals are born than can possibly survive. Thus, for example, it has been calculated that if the progeny of a single pair of elephants--which are the slowest breeding of animals--were all allowed to reach maturity and propagate, in 750 years there would be living 19,000,000 descendants. Again, in the case of vegetables, if a species of annual plant produces only two seeds a year, if these in successive years were all allowed to reproduce their kind, in twenty years there would be 11,000,000 plants from a single ancestor. Yet we know that nearly all animals and plants produce many more young at a time than in either of these two supposed cases. Indeed, as individuals of many kinds of plants, and not a few kinds of animals, produce every year several thousand young, we may make a rough estimate and say, that over organic nature as a whole probably not one in a thousand young are allowed to survive to the age of reproduction. How tremendous, therefore, must be the struggle for existence! It is thought a terrible thing in battle when one half the whole number of combatants perish. But what are we to think of a battle for life where only one in a thousand survives?

This, then, is the first fact. The second is the fact so long ago recognised, that the battle is to the strong, the race to the swift. The thousandth individual which does survive in the battle for existence--which does win the race for life--is, without question, one of the individuals best fitted to do so; that is to say, best fitted to the conditions of its existence considered as a whole. Nature is, therefore, always picking out, or selecting, such individuals to live and to breed.

The third fact is, that the individuals so selected transmit their favourable qualities to their offspring by heredity. There is no doubt about this fact, so far as we are concerned with it. For although, as I have already hinted, considerable doubt has of late years been cast upon Lamarck's doctrine of the hereditary transmission of acquired characters, it remains as impossible as ever it was to question the hereditary transmission of what are called congenital characters. And this is all that Darwin's theory necessarily requires.

The fourth fact is, that although heredity as a whole produces a wonderfully exact copy of the parent in the child, there is never a precise reduplication. Of all the millions of human beings upon the face of the earth, no one is so like another that we cannot see some difference; the resemblance is everywhere specific, nowhere individual. Now this same remark applies to all specific types. The only reason why we notice individual differences in the case of the human type more than we do in the case of any other types, is because our attention is here more incessantly focussed upon these differences. We are compelled to notice them in the case of our own species, however small they may appear to a naturalist, because, unless we do so, we should not recognise the members of our own family, or be able to distinguish between a man whom we know is ready to do us an important service, and another man whom we know is ready to cut our throats. But our common mother Nature is able thus to distinguish between all her children. Her eyes are much more ready to detect small individual peculiarities than are the eyes of any naturalist. No slight variations in the cast of feature or disposition of parts, no minute difference in the arrangement of microscopical cells, can escape her ever vigilant attention. And, consequently, when among all the innumerable multitudes of individual variations any one arises which--no matter in how slight a degree--gives to that individual a better chance of success in the struggle for life, Nature chooses that individual to survive, and so to perpetuate the improvement in his or her progeny.

Now I say that all these several component parts of Darwinian doctrine are not matters of theory, but matters of fact. The only element of theory in his doctrine of evolution by natural selection has reference to the degree in which these observable facts, when thus brought together, are adequate to account for the process of evolution.

* * * * *

So much, then, as a statement of the theory of natural selection. But from this statement--i. e. from the theory of natural selection itself--there follow certain matters of general principle which it is important to bear in mind. These, therefore, I shall here proceed to mention.

First of all, it is evident that the theory is applicable as an explanation of organic changes in specific types only in so far as these changes are of use, or so far as such changes endow the species with better chances of success in the general struggle for existence. This is the only sense in which I shall always employ the terms use, utility, service, benefit, and so forth--that is to say, in the sense of life-preserving.

* * * * *

Next, it must be clearly understood that the life which it is the object, so to speak, of natural selection to preserve, is primarily the life of the species; not that of the individual. Natural selection preserves the life of the individual only in so far as this is conducive to that of the species. Wherever the life-interests of the individual clash with those of the species, that individual is sacrificed in favour of others who happen better to subserve the interests of the species. For example, in all organisms a greater or less amount of vigour is wasted, so far as individual interests are concerned, in the formation and the nourishment of progeny. In the great majority of plants and animals an enormous amount of physiological energy is thus expended. Look at the roe or the milt of a herring, for instance, and see what a huge drain has been made upon the individual for the sake of its species. Again, all unselfish instincts have been developed for the sake of the species, and usually against the interests of the individual. An ant which will allow her head to be slowly drawn from her body rather than relinquish her hold upon a pupa, is clearly acting in response to an instinct which has been developed for the benefit of the hive, though fatal to the individual. And, in a lesser degree, the parental instincts, wherever they occur, are more or less detrimental to the interests of the individual, though correspondingly essential to those of the race.

These illustrations will serve to show that natural selection always works primarily for the life-interests of the species--and, indeed, only works for those of the individual at all in so far as the latter happen to coincide with the former. Or, otherwise stated, the object of natural selection is always that of producing and maintaining specific types in the highest degree of efficiency, no matter what may become of the constituent individuals. Which is a striking republication by Science of a general truth previously stated by Poetry:--

So careful of the type she seems, So careless of the single life.

Tennyson thus noted the fact, and a few years later Darwin supplied the explanation.

But of course in many, if not in the majority of cases, anything that adds to the life-sustaining power of the single life thereby ministers also to the life-sustaining power of the type; and thus we can understand why all mechanisms and instincts which minister to the single life have been developed--namely, because the life of the species is made up of the lives of all its constituent individuals. It is only where the interests of the one clash with those of the other that natural selection works against the individual. So long as the interests are coincident, it works in favour of both.

Natural selection, then, is a theory which seeks to explain by natural causes the occurrence of every kind of adaptation which is to be met with in organic nature, on the assumption that adaptations of every kind have primary reference to the preservation of species, and therefore also, as a general rule, to the preservation of their constituent individuals. And from this it follows that where it is for the benefit of a species to change its type, natural selection will effect that change, thus leading to a specific transmutation, or the evolution of a new species. In such cases the old species may or may not become extinct. If the transmutation affects the species as a whole, or throughout its entire range, of course that particular type becomes extinct, although it does so by becoming changed into a still more suitable type in the course of successive generations. If, on the other hand, the transmutation affects only a part of the original species, or not throughout its entire range, then the other parts of that species may survive for any number of ages as they originally were. In the one case there is a ladder-like transmutation of species in time; in the other case a possibly tree-like multiplication of species in space. But whether the evolution of species be thus serial in time or divergent in space, the object of natural selection, so to speak, is in either case the same--namely, that of preserving all types which prove best suited to the conditions of their existence.

* * * * *

Once more, the term "struggle for existence" must be understood to comprehend, not only a competition for life among contemporary individuals of the same species, but likewise a struggle by all such individuals taken collectively for the continuance of their own specific type. Thus, on the one hand, while there is a perpetual civil war being waged between members of the same species, on the other hand there is a foreign war being waged by the species as a whole against its world as a whole. Hence it follows that natural selection does not secure survival of the fittest as regards individuals only, but also survival of the fittest as regards types. This is a most important point to remember, because, as a general rule, these two different causes produce exactly opposite effects. Success in the civil war, where each is fighting against all, is determined by individual fitness and self-reliance. But success in the foreign war is determined by what may be termed tribal fitness and mutual dependence. For example, among social insects the struggle for existence is quite as great between different tribes or communities, as it is between different individuals of the same community; and thus we can understand the extraordinary degree in which not only co-operative instincts, but also largely intelligent social habits, have here been developed[30]. Similarly, in the case of mankind, we can understand the still more extraordinary development of these things--culminating in the moral sense. I have heard a sermon, preached at one of the meetings of the British Association, entirely devoted to arguing that the moral sense could not have been evolved by natural selection, seeing that the altruism which this sense involves is the very opposite of selfishness, which alone ought to have been the product of survival of the fittest in a struggle for life. And, of course, this argument would have been perfectly sound had Darwin limited the struggle for existence to individuals, without extending it to communities. But if the preacher had ever read Darwin's works he would have found that, when thus extended, the principle of natural selection is bound to work in favour of the co-operative instincts in the case of so highly social an animal as man; and that of these instincts conscience is the highest imaginable exhibition.

[30] For cases, see Animal Intelligence, in the chapters on Ants and Bees; and, for discussion of principles, Mental Evolution in Animals, in the chapters on Instinct.

What I have called tribal fitness--in contradistinction to individual fitness-- begins with the family, developes in the community (herd, hive, clan, &c.), and usually ends with the limits of the species. On the one hand, however, it is but seldom that it extends so far as to embrace the entire species; while, on the other hand, it may in some cases, and as it were sporadically, extend beyond the species. In these latter cases members of different species mutually assist one another, whether in the way of what is called symbiosis, or in a variety of other ways which I need not wait to mention. For the only point which I now desire to make clear is, that all cases of mutual aid or co-operation, whether within or beyond the limits of species, are cases which fall under the explanatory sweep of the Darwinian theory[31].

[31] Prince Kropotkin in the Nineteenth Century (Feb. 1888, Apr. 1891) has adduced a large and interesting body of facts, showing the great prevalence of the principle of co-operation in organic nature.

* * * * *

Another important point to notice is, that it constitutes no part of the theory of natural selection to suppose that survival of the fittest must invariably lead to improvement of type, in the sense of superior organization. On the contrary, if from change of habits or conditions of life an organic type ceases to have any use for previously useful organs, natural selection will not only allow these organs in successive generations to deteriorate--by no longer placing any selective premium upon their maintenance--but may even proceed to assist the agencies engaged in their destruction. For, being now useless, they may become even deleterious, by absorbing nutriment, causing weight, occupying space, &c., without conferring any compensating benefit. Thus we can understand why it is that parasites, for example, present the phenomena of what is called degeneration, i. e. showing by their whole structure that they have descended from a possibly very much higher type of organization than that which they now exhibit. Having for innumerable generations ceased to require their legs, their eyes, and so forth, all such organs of high elaboration have either disappeared or become vestigial, leaving the parasite as a more or less effete representative of its ancestry.

These facts of degeneration, as we have previously seen, are of very general occurrence, and it is evident that their importance in the field of organic evolution as a whole has been very great. Moreover, it ought to be particularly observed that, as just indicated, the facts may be due either to a passive cessation of selection, or to an active reversal of it. Or, more correctly, these facts are probably always due to the cessation of selection, although in most cases where species in a state of nature are concerned, the process of degeneration has been both hastened and intensified by the super-added influence of the reversal of selection. In the next volume I shall have occasion to recur to this distinction, when it will be seen that it is one of no small importance to the general theory of descent.

* * * * *

We may now proceed to consider certain misconceptions of the Darwinian theory which are largely, not to say generally, prevalent among supporters of the theory. These misconceptions, therefore, differ from those which fall to be considered in the next chapter, i. e. misconceptions which constitute grounds of objection to the theory.

* * * * *

Of all the errors connected with the theory of natural selection, perhaps the one most frequently met with--especially among supporters of the theory--is that of employing the theory to explain all cases of phyletic modification (or inherited change of type) indiscriminately, without waiting to consider whether in particular cases its application is so much as logically possible. The term "natural selection" thus becomes a magic word, or Sesame, at the utterance of which every closed door is supposed to be immediately opened. Be it observed, I am not here alluding to that merely blind faith in natural selection, which of late years has begun dogmatically to force this principle as the sole cause of organic evolution in every case where it is logically possible that the principle can have come into play. Such a blind faith, indeed, I hold to be highly inimical, not only to the progress of biological science, but even to the true interests of the natural selection theory itself. As to this I shall have a good deal to say in the next volume. Here, however, the point is, that the theory in question is often invoked in cases where it is not even logically possible that it can apply, and therefore in cases where its application betokens, not merely an error of judgment or extravagance of dogmatism, but a fallacy of reasoning in the nature of a logical contradiction. Almost any number of examples might be given; but one will suffice to illustrate what is meant. And I choose it from the writings of one of the authors of the selection theory itself, in order to show how easy it is to be cheated by this mere juggling with a phrase--for of course I do not doubt that a moment's thought would have shown the writer the untenability of his statement.

In his most recent work Mr. Wallace advances an interesting hypothesis to the effect that differences of colour between allied species, which are apparently too slight to serve any other purpose, may act as "recognition marks," whereby the opposite sexes are enabled at once to distinguish between members of their own and of closely resembling species. Of course this hypothesis can only apply to the higher animals; but the point here is that, supposing it to hold for them, Mr. Wallace proceeds to argue thus:-- Recognition marks "have in all probability been acquired in the process of differentiation for the purpose of checking the intercrossing of allied forms," because "one of the first needs of a new species would be to keep separate from its nearest allies, and this could be more readily done by some easily seen external mark[32]." Now, it is clearly not so much as logically possible that these recognition-marks (supposing them to be such) can have been acquired by natural selection, "for the purpose of checking intercrossing of allied forms." For the theory of natural selection, from its own essential nature as a theory, is logically exclusive of the supposition that survival of the fittest ever provides changes in anticipation of future uses. Or, otherwise stated, it involves a contradiction of the theory itself to say that the colour-changes in question were originated by natural selection, in order to meet "one of the first needs of a new species," or for the purpose of subsequently preventing intercrossing with allied forms. If it had been said that these colour-differentiations were originated by some cause other than natural selection (or, if by natural selection, still with regard to some previous, instead of prophetic, "purpose"), and, when so "acquired," then began to serve the "purpose" assigned, the argument would not have involved the fallacy which we are now considering. But, as it stands, the argument reverts to the teleology of pre-Darwinian days--or the hypothesis of a "purpose" in the literal sense which sees the end from the beginning, instead of a "purpose" in the metaphorical sense of an adaptation that is evolved by the very modifications which subserve it[33].

[32] Darwinism, pp. 218 and 227.

[33] Since the above was written Prof. Lloyd Morgan has published a closely similar notice of the passage in question. "This language," he says, "seems to savour of teleology (that pitfall of the evolutionist). The cart is put before the horse. The recognition-marks were, I believe, not produced to prevent

intercrossing, but intercrossing has been prevented because of preferential mating between individuals possessing special recognition-marks. To miss this point is to miss an important segregation-factor."--(Animal Life and Intelligence, p. 103.) Again, on pp. 184-9, he furnishes an excellent discussion on the whole subject of the fallacy alluded to in the text, and gives illustrative quotations from other prominent Darwinians. I should like to add that Darwin himself has nowhere fallen into this, or any of the other fallacies, which are mentioned in the text.

* * * * *

Another very prevalent, and more deliberate, fallacy connected with the theory of natural selection is, that it follows deductively from the theory itself that the principle of natural selection must be the sole means of modification in all cases where modification is of an adaptive kind,--with the consequence that no other principle can ever have been concerned in the production of structures or instincts which are of any use to their possessors. Whether or not natural selection actually has been the sole means of adaptive modification in the race, as distinguished from the individual, is a question of biological fact[34]; but it involves a grave error of reasoning to suppose that this question can be answered deductively from the theory of natural selection itself, as I shall show at some length in the next volume.

[34] Of course adaptive modifications produced in the individual lifetime, and not inherited, do not concern the question at all. In this and the following paragraphs, therefore, "adaptations," "adaptive modifications," &c., refer exclusively to such as are hereditary, i. e. phyletic.

* * * * *

A still more extravagant, and a still more unaccountable fallacy is the one which represents it as following deductively from the theory of natural selection itself, that all hereditary characters are "necessarily" due to natural selection. In other words, not only all adaptive, but likewise all non-adaptive hereditary characters, it is said, must be due to natural selection. For non-adaptive characters are taken to be due to "correlation of growth," in connexion with some of the adaptive ones--natural selection being thus the indirect means of producing the former wherever they may occur, on account of its being the direct and the only means of producing the latter. Thus it is deduced from the theory of natural selection itself,--1st, that the principle of natural selection is the only possible cause of adaptive modification: 2nd, that non-adaptive modifications can only occur in the race as correlated appendages to the adaptive: 3rd, that, consequently, natural selection is the only possible cause of modification, whether adaptive or non-adaptive. Here again, therefore, we must observe that none of these sweeping generalizations can possibly be justified by deductive reasoning from the theory of natural selection itself. Any attempt at such deductive reasoning must necessarily end in circular reasoning, as I shall likewise show in the second volume, where this whole "question of utility" will be thoroughly dealt with.

* * * * *

Once more, there is an important oversight very generally committed by the followers of Darwin. For even those who avoid the fallacies above mentioned often fail to perceive, that natural selection can only begin to operate if the degree of adaptation is already given as sufficiently high to count for something in the struggle for existence. Any adaptations which fall below this level of importance cannot possibly have been produced by survival of the fittest. Yet the followers of Darwin habitually speak of adaptative characters, which in their own opinion are subservient merely to comfort or convenience, as having been produced by such means. Clearly this is illogical; for it belongs to the essence of Darwin's theory to suppose, that natural selection can have no jurisdiction beyond the line where structures or instincts already present a sufficient degree of adaptational value to increase, in some measure, the expectation of life on the part of their possessors. We cannot speak of adaptations as due to natural selection, without thereby affirming that they present what I have elsewhere termed a "selection value."

* * * * *

Lastly, as a mere matter of logical definition, it is well-nigh self-evident that the theory of natural selection is a theory of the origin, and cumulative development, of adaptations, whether these be distinctive of species, or of genera, orders, families, classes, and sub-kingdoms. It is only when the adaptations happen to be distinctive of the first (or lowest) of these taxonomic divisions, that the theory which accounts for these adaptations accounts also for the forms which present them,--i. e. becomes also a theory of the origin of species. This, however, is clearly but an accident of particular cases; and, therefore, even in them the theory is primarily a theory of adaptations, while it is but secondarily a theory of the species which present them. Or, otherwise stated, the theory is no more a theory of the origin of species than it is of the origin of genera, families, and the rest; while, on the other hand, it is everywhere a theory of the adaptive modifications whereby each of these taxonomic divisions has been differentiated as such. Yet, sufficiently obvious as the accuracy of this definition must appear to any one who dispassionately considers it, several naturalists of high standing have denounced it in violent terms. I shall therefore have to recur to the subject at somewhat greater length hereafter. At present it is enough merely to mention the matter, as furnishing another and a curious illustration of the not infrequent weakness of logical perception on the part of minds well gifted with the faculty of observation. It may be added, however, that the definition in question is in no way hostile to the one which is virtually given by Darwin in the title of his great work. The Origin of Species by means of Natural Selection is beyond doubt the best title that could have been given, because at the time when the work was published the fact, no less than the method, of organic evolution had to be established; and hence the most important thing to be done at that time was to prove the transmutation of species. But now that this has been done to the satisfaction of naturalists in general, it is as I have said, curious to find some of them denouncing a wider definition of the principle of natural selection, merely because the narrower (or included) definition is invested with the charm of verbal associations[35].

[35] The question as to whether natural selection has been the only principle concerned in the origination of species, is quite distinct from that as to the accuracy of the above definition.

* * * * *

So much for fallacies and misconceptions touching Darwin's theory, which are but too frequently met with in the writings of its supporters. We must now pass on to mention some of the still greater fallacies and misconceptions which are prevalent in the writings of its opponents. And, in order to do this thoroughly, I shall begin by devoting the remainder of the present chapter to a consideration of the antecedent standing of the two theories of natural selection and supernatural design. This having been done, in the succeeding chapters I shall deal with the evidences for, and the objections against, the former theory.

* * * * *

Beginning, then, with the antecedent standing of these alternative theories, the first thing to be noticed is, that they are both concerned with the same subject-matter, which it is their common object to explain. Moreover, this subject-matter is clearly and sharply divisible into two great classes of facts in organic nature--namely, those of Adaptation and those of Beauty. Darwin's theory of descent explains the former by his doctrine of natural selection, and the latter by his doctrine of sexual selection. In the first instance, therefore, I shall have to deal only with the facts of adaptation, leaving for subsequent consideration the facts of beauty.

Innumerable cases of the adaptation of organisms to their surroundings being the facts which now stand before us to be explained either by natural selection or by supernatural intention, we may first consider a statement which is frequently met with--namely, that even if all such cases of adaptation were proved to be fully explicable by the theory of descent, this would constitute no disproof of the theory of design: all the cases of adaptation, it is argued, might still be due to design, even though they admit of being hypothetically accounted for by the theory of descent. I have heard an eminent Professor tell his class that the many instances of mechanical adaptation discovered and described by Darwin as occurring in orchids, seemed to him to furnish better proof of supernatural contrivance than of natural causes; and another eminent Professor has informed me that, although he had read the Origin of Species with care, he could see in it no evidence of natural selection which might not equally well have been adduced in favour of intelligent design. But here we meet with a radical misconception of the whole logical attitude of science. For, be it observed, this exception in limine to the evidence which we are about to consider does not question that natural selection may be able to do all that Darwin ascribes to it. The objection is urged against his interpretation of the facts merely on the ground that these facts might equally well be ascribed to intelligent design. And so undoubtedly they might, if we were all simple enough to adopt a supernatural explanation whenever a natural one is found sufficient to account for the facts. Once admit the irrational principle that we may assume the operation of higher causes where the operation of lower ones is sufficient to explain the observed phenomena, and all our science and all our philosophy are scattered to the winds. For the law of logic which Sir William Hamilton called the law of parsimony--or the law which forbids us to assume the operation of higher causes when lower ones are found sufficient to explain the observed effects--this law constitutes the only barrier between science and superstition. It is always possible to give a hypothetical explanation of any phenomenon whatsoever, by referring it immediately to the intelligence of some supernatural agent; so that the only difference between the logic of science and the logic of superstition consists in science recognising a validity in the law of parsimony which superstition disregards. Therefore one can have no hesitation in saying that this way of looking at the evidence in favour of natural selection is not a scientific or a reasonable way of looking at it, but a purely superstitious way. Let us take, as an illustration, a perfectly parallel case. When Kepler was unable to explain by any known causes the paths described by the planets, he resorted to a supernatural explanation, and supposed that every planet was guided in its movements by some presiding angel. But when Newton supplied a beautifully simple physical explanation, all persons with a scientific habit of mind at once abandoned the metaphysical one. Now, to be consistent, the above-mentioned Professors, and all who think with them, ought still to adhere to Kepler's hypothesis in preference to Newton's explanation; for, excepting the law of parsimony, there is certainly no other logical objection to the statement, that the movements of the planets afford as good evidence of the influence of guiding angels as they do of the influence of gravitation.

So much, then, for the illogical position that, granting the evidence in favour of natural descent and supernatural design to be equal and parallel, we should hesitate in our choice between the two theories. But, of course, if the evidence is supposed not to be equal and parallel--i. e. if it is supposed that the theory of natural selection is not so good a theory whereby to explain the facts of adaptation as is that of supernatural design,--then the objection is no longer the one which we are considering. It is quite another objection, and one which is not prima facie absurd. Therefore let us state clearly the distinct question which thus arises.

Innumerable cases of adaptation of organisms to their environments are the observed facts for which an explanation is required. To supply this explanation, two, and only two, hypotheses are in the field. Of these two hypotheses one is intelligent design manifested directly in special creation; the other is natural causation operating through countless ages of the past. Now, the adaptations in question involve an innumerable multitude of special mechanisms, in most cases even within the limits of any one given species; but when we consider the sum of all these mechanisms presented by organic nature as a whole, the mind must indeed be dull which does not feel astounded. For, be it further observed, these mechanical contrivances[36] are, for the most part, no merely simple arrangements, which might reasonably be supposed due, like the phenomena of crystallization, to comparatively simple physical causes. On the contrary, they everywhere and habitually exhibit so deep-laid, so intricate, and often so remote an adaptation of means to ends, that no machinery of human contrivance can properly be said to equal their perfection from a mechanical point of view. Therefore, without question, the hypothesis which first of all they suggest--or suggest most readily--is the hypothesis of design. And this hypothesis becomes virtually the only hypothesis possible, if it be assumed--as it generally was assumed by natural theologians of the past,--that all species of plants and animals were introduced into the world suddenly. For it is quite inconceivable that any known cause, other than intelligent design, could be competent to turn out instantaneously any one of these intricate pieces of machinery, already adapted to the performance of its special function. But, on the other hand, if there is any evidence to show that one species becomes slowly transformed into another--or that one set of adaptations becomes slowly changed into another set as changing circumstances require,--then it becomes quite possible to imagine that a strictly natural causation may have had something to do with the matter. And this suggestion becomes greatly more probable when we discover, from geological evidence and embryological research, that in the history both of races and of individuals the various mechanisms in question have themselves had a history--beginning in the forms of most uniformity and simplicity, gradually advancing to forms more varied and complex, nowhere exhibiting any interruptions in their upward progress, until the world of organic machinery as we now have it is seen to have been but the last phase of a long and gradual growth, the ultimate roots of which are to be found in the soil of undifferentiated protoplasm.

[36] It is often objected to Darwin's terminology, that it embraces such words as "contrivance," "purpose," &c., which are strictly applicable only to the processes or the products of thought. But when it is understood that they are used in a neutral or metaphorical sense, I cannot see that any harm arises from their use.

Lastly, when there is supplied to us the suggestion of natural selection as a cause presumably adequate to account for this continuous growth in the number, the intricacy, and the perfection of such mechanisms, it is only the most unphilosophical mind that can refuse to pause as between the older hypothesis of design and the newer hypothesis of descent.

Thus it is clear that the a priori standing of the rival hypotheses of naturalism and supernaturalism in the case of all these pieces of organic machinery, is profoundly affected by the question whether they came into existence suddenly, or whether they did so gradually. For, if they all came into existence suddenly, the fact would constitute well-nigh positive proof in favour of supernaturalism, or creation by design; whereas, if they all came into existence gradually, this fact would in itself constitute presumptive evidence in favour of naturalism, or of development by natural causes. And, as shown in the previous chapters, the proof that all species of plants and animals came into existence gradually--or the proof of evolution as a fact--is simply overwhelming.

From a still more general point of view I may state the case in another way, by borrowing and somewhat expanding an illustration which, I believe, was first used by Professor Huxley. If, when the tide is out, we see lying upon the shore a long line of detached sea-weed, marking the level which is reached by full tide, we should be free to conclude that the separation of the sea-weed from the sand and the stones was due to the intelligent work of some one who intended to collect the sea-weed for manure, or for any other purpose. But, on the other hand, we might explain the fact by a purely physical cause-- namely, the separation by the sea-waves of the sea-weed from the sand and stones, in virtue of its lower specific gravity. Now, thus far the fact would be explained equally well by either hypothesis; and this fact would be the fact of selection. But whether we yielded our assent to the one explanation or to the other would depend upon a due consideration of all collateral circumstances. The sea-weed might not be of a kind that is of any use to man; there might be too great a quantity of it to admit of our supposing that it had been collected by man; the fact that it was all deposited on the high-water-mark would in itself be highly suggestive of the agency of the sea; and so forth. Thus, in such a case any reasonable observer would decide in favour of the physical explanation, or against the teleological one.

Now the question whether organic evolution has been caused by physical agencies or by intelligent design is in precisely the same predicament. There can be no logical doubt that, theoretically at all events, the physical agencies which the present chapter is concerned with, and which are conveniently summed up in the term natural selection, are as competent to produce these so-called mechanical contrivances, and the other cases of adaptation which are to be met with in organic nature, as intelligent design could be. Hence, our choice as between these two hypotheses must be governed by a study of all collateral circumstances; that is to say, by a study of the evidences in favour of the physical explanation. To this study, therefore, we shall now address ourselves, in the course of the following chapters.

CHAPTER VIII.

EVIDENCES OF THE THEORY OF NATURAL SELECTION.

I will now proceed to state the main arguments in favour of the theory of natural selection, and then, in the following chapter, the main objections which have been urged against it.

In my opinion, the main arguments in favour of the theory are three in number.

First, it is a matter of observation that the struggle for existence in nature does lead to the extermination of forms less fitted for the struggle, and thus makes room for forms more fitted. This general fact may be best observed in cases where an exotic species proves itself better fitted to inhabit a new country than is some endemic species which it exterminates. In Great Britain, for example, the so-called common rat is a comparatively recent importation from Norway, and it has so completely supplanted the original British rat, that it is now extremely difficult to procure a single specimen of the latter: the native black rat has been all but exterminated by the foreign brown rat. The same thing is constantly found in the case of imported species of plants. I have seen the river at Cambridge so choked with the inordinate propagation of a species of water-weed which had been introduced from America, that considerable expense had to be incurred in order to clear the river for traffic. In New Zealand the same thing has happened with the European water-cress, and in Australia with the common rabbit. So it is doubtless true, as one of the natives is said to have philosophically remarked, "the white man's rat has driven away our rat, the European fly drives away our fly, his clover kills our grass, and so will the Maoris disappear before the white man himself." Innumerable other cases to the same effect might be quoted; and they all go to establish the fact that forms less fitted to survive succumb in their competition with forms better fitted.

* * * * *

Secondly, there is a general consideration of the largest possible significance in the present connexion--namely, that among all the millions of structures and instincts which are so invariably, and for the most part so wonderfully, adapted to the needs of the species presenting them, we cannot find a single instance, either in the vegetable or animal kingdom, of a structure or an instinct which is developed for the exclusive benefit of another species. Now this great and general fact is to my mind a fact of the most enormous, not to say overwhelming, significance. The theory of natural selection has now been before the world for more than thirty years, and during that time it had stood a fire of criticism such as was never encountered by any scientific theory before. From the first Darwin invited this criticism to adduce any single instance, either in the vegetable or animal kingdom, of a structure or an instinct which should unquestionably be proved to be of exclusive use to any species other than the one presenting it. He even went so far as to say that if any one such instance could be shown he would surrender his whole theory on the strength of it--so assured had he become, by his own prolonged researches, that natural selection was the true agent in the production of adaptive structures, and, as such, could never have permitted such a structure to occur in one species for the benefit of another. Now, as this invitation has been before the world for so many years, and has not yet been answered by any naturalist, we may by this time be pretty confident that it never will be answered. How tremendous, then, is the significance of this fact in its testimony to Darwin's theory! The number of animal and vegetable species, both living and extinct, is to be reckoned by millions, and every one of these species presents on an average hundreds of adaptive structures,--at least one of which in many, possibly in most, if not actually in all cases, is peculiar to the species that presents it. In other words, there are millions of adaptive structures (not to speak of instincts) which are peculiar to the species presenting them, and also many more which are the common property of allied species: yet, notwithstanding this inconceivable profusion of adaptive structures in organic nature, there is no single instance that has been pointed out of the occurrence of such a structure save for the benefit of the species that presents it. Therefore, I say that this immensely large and general fact speaks with literally immeasurable force in favour of natural selection, as at all events one of the main causes of organic evolution. For the fact is precisely what we should expect if this theory is true, while upon no other theory can its universality and invariability be rendered intelligible. On the beneficent design theory, for instance, it is inexplicable that no species should ever be found to present a structure or an instinct having primary reference to the welfare of another species, when, ex hypothesi, such an endless amount of thought has been displayed in the creation of structures and instincts having primary reference to the species which present them. For how magnificent a display of divine beneficence would organic nature have afforded, if all--or even some--species had been so inter-related as to have ministered to each others wants. Organic species might then have been likened to a countless multitude of voices, all singing in one great harmonious psalm. But, as it is, we see absolutely no vestige of such co-ordination: every species is for itself, and for itself alone--an outcome of the always and everywhere fiercely raging struggle for life.

In order that the force of this argument may not be misapprehended, it is necessary to bear in mind that it is in no way affected by cases where a structure or an instinct is of primary benefit to its possessor, and then becomes of secondary benefit to some other species on account of the latter being able in some way or another to utilise its action. Of course organic nature is full of cases of this kind; but they only go to show the readiness which all species display to utilise for themselves everything that can be turned to good account in their own environments, and so, among other things, the structures and instincts of other animals. For instance, it would be no answer to Darwin's challenge if any one were to point to a hermit-crab inhabiting the cast-off shell of a mollusk; because the shell was primarily of use to the mollusk itself, and, so far as the mollusk is concerned, the fact of its shell being afterwards of a secondary use to the crab is quite immaterial. What Darwin's challenge requires is, that some structure or instinct should be shown which is not merely of such secondary or accidental benefit to another species, but clearly adapted to the needs of that other species in the first instance--such, for example, as would be the case if the tail of a rattle-snake were of no use to its possessor, while serving to warn other animals of the proximity of a dangerous creature; or, in the case of instincts, if it were true that a pilot-fish accompanies a shark for the purpose of helping the shark to discover food. Both these instances have been alleged; but both have been shown untenable. And so it has proved of all the other cases which thus far have been put forward.

Perhaps the most remarkable of all the allegations which ever have been put forward in this connexion are those that were current with regard to instincts before the publication of Darwin's work. These allegations are the most remarkable, because they serve to show, in a degree which I do not believe could be shown anywhere else, the warping power of preconceived ideas. A short time ago I happened to come across the 8th edition of the Encyclop 鎬 ia Britannica, and turned up the article on "Instinct" there, in order to see what amount of change had been wrought with regard to our views on this subject by the work of Darwin--the 8th edition of the Encyclop 鎬 ia Britannica having been published shortly before The Origin of Species by means of Natural Selection. I cannot wait to give any lengthy quotations from this representative exponent of scientific opinion upon the subject at that time; but its general drift may be appreciated if I transcribe merely the short concluding paragraph, wherein he sums up his general results. Here he says:--

It thus only remains for us to regard instinct as a mental faculty, sui generis, the gift of God to the lower animals, that man in his own person, and by them, might be relieved from the meanest drudgery of nature.

Now, here we have the most extraordinary illustration that is imaginable of the obscuring influence of a preconceived idea. Because he started with the belief that instincts must have been implanted in animals for the benefit of man, this writer, even when writing a purely scientific essay, was completely blinded to the largest, the most obvious, and the most important of the facts which the phenomena of instinct display. For, as a matter of fact, among all the many thousands of instincts which are known to occur in animals, there is no single one that can be pointed to as having any special reference to man; while, on the other hand, it is equally impossible to point to one which does not refer to the welfare of the animal presenting it. Indeed, when the point is suggested, it seems to me surprising how few in number are the instincts of animals which have proved to be so much as of secondary or accidental benefit to man, in the same way as skins, furs, and a whole host of other animal products are thus of secondary use to him. Therefore, this writer not only failed to perceive the most obvious truth that every instinct, without any single exception, has reference to the animal which presents it; but he also conceived a purely fictitious inversion of this truth, and wrote an essay to prove a statement which all the instincts in the animal kingdom unite in contradicting.

This example will serve to show, in a striking manner, not only the distance that we have travelled in our interpretation of organic nature between two successive editions of the Encyclop鑗ia Britannica, but also the amount of verification which this fact furnishes to the theory of natural selection. For, inasmuch as it belongs to the very essence of this theory that all adaptive characters (whether instinctive or structural) must have reference to their own possessors, we find overpowering verification furnished to the theory by the fact now before us--namely, that immediately prior to the enunciation of this theory, the truth that all adaptive characters have reference only to the species which present them was not perceived. In other words, it was the testing of this theory by the facts of nature that revealed to naturalists the general law which the theory, as it were, predicted--the general law that all adaptive characters have primary reference to the species which present them. And when we remember that this is a kind of verification which is furnished by millions of separate cases, the whole mass of it taken together is, as I have before said, overwhelming.

It is somewhat remarkable that the enormous importance of this argument in favour of natural selection as a prime factor of organic evolution has not received the attention which it deserves. Even Darwin himself, with his characteristic reserve, has not presented its incalculable significance; nor do I know any of his followers who have made any approach to an adequate use of it in their advocacy of his views. In preparing the present chapter, therefore, I have been particularly careful not to pitch too high my own estimate of its evidential value. That is to say, I have considered, both in the domain of structures and of instincts, what instances admit of being possibly adduced per contra, or as standing outside the general law that adaptive structures and instincts are of primary use only to their possessors. In the result I can only think of two such instances. These, therefore, I will now dispose of.

The first was pointed out, and has been fully discussed, by Darwin himself. Certain species of ants are fond of a sweet fluid that is secreted by aphides, and they even keep the aphides as we keep cows for the purpose of profiting by their "milk." Now the point is, that the use of this sweet secretion to the aphis itself has not yet been made out. Of course, if it is of no use to the aphis, it would furnish a case which completely meets Darwin's own challenge. But, even if this supposition did not stand out of analogy with all the other facts of organic nature, most of us would probably deem it prudent to hold that the secretion must primarily be of some use to the aphis itself, although the matter has not been sufficiently investigated to inform us of what this use is. For, in any case, the secretion is not of any vital importance to the ants which feed upon it: and I think but few impartial minds would go so far to save an hypothesis as to maintain, that the Divinity had imposed this drain upon the internal resources of one species of insect for the sole purpose of supplying a luxury to another. On the whole, it seems most probable that the fluid is of the nature of an excretion, serving to carry off waste products. Such, at all events, was the opinion at which Darwin himself arrived, as a result of observing the facts anew, and in relation to his theory.

* * * * *

The other instance to which I have alluded as seeming at first sight likely to answer Darwin's challenge is the formation of vegetable galls. The great number and variety of galls agree in presenting a more or less elaborate structure, which is not only foreign to any of the uses of plant-life, but singularly and specially adapted to those of the insect-life which they shelter. Yet they are produced by a growth of the plant itself, when suitably stimulated by the insects' inoculation--or, according to recent observations, by emanations from the bodies of the larv?which develop from the eggs deposited in the plant by the insect. Now, without question, this is a most remarkable fact; and if there were many more of the like kind to be met with in organic nature, we might seriously consider whether the formation of galls should not be held to make against the ubiquitous agency of natural selection. But inasmuch as the formation of galls stands out as an exception to the otherwise universal rule of every species for itself, and for itself alone, we are justified in regarding this one apparent exception with extreme suspicion. Indeed, I think we are justified in regarding the peculiar pathological effect produced in the plant by the secretions of the insect as having been in the first instance accidentally beneficial to the insects. Thus, if any other effect than that of a growing tumour had been produced in the first instance, or if the needs of the insect progeny had not been such as to have derived profit from being enclosed in such a tumour, then, of course, the inoculating instinct of these animals could not have been developed by natural selection. But, given these two conditions, and it appears to me there is nothing very much more remarkable about an accidental correlation between the effects of a parasitic larva on a plant and the needs of that parasite, than there is between the similarly accidental correlation between a hydated parasite and the nutrition furnished to it by the tissues of a warm-blooded animal. Doubtless the case of galls is somewhat more remarkable, inasmuch as the morbid growth of the plant has more concern in the correlation--being, in many instances, a more specialized structure on the part of a host than occurs anywhere else, either in the animal or vegetable world. But here I may suggest that although natural selection cannot have acted upon the plant directly, so as to have produced galls ever better and better adapted to the needs of the insect, it may have so acted upon the plants indirectly though the insects. For it may very well have been that natural selection would ever tend to preserve those individual insects, the quality of whose emanations tended to produce the form of galls best suited to nourish the insect progeny; and thus the character of these pathological growths may have become ever better and better adapted to the needs of the insects. Lastly, looking to the enormous number of relations and inter-relations between all organic species, it is scarcely to be wondered at that even so extraordinary an instance of correlation as this should have arisen thus by accident, and then have been perfected by such an indirect agency of natural selection as is here suggested[37].

[37] Note B.

* * * * *

The third general class of facts which tell so immensely in favour of natural selection as an important cause of organic evolution, are those of domestication. The art of the horticulturist, the fancier, the cattle-breeder, &c., consists in producing greater and greater deviations from a given wild type of plant or animal, in any particular direction that may be desired for purposes either of use or of beauty. Cultivated cereals, fruits, and flowers are known to have been all derived from wild species; and, of course, the same applies to all our domesticated varieties of animals. Yet if we compare a cabbage rose with a wild rose, a golden pippin apple with a crab, a toy terrier with any species of wild dog, not to mention any number of other instances, there can be no question that, if such differences had appeared in nature, the organisms presenting them would have been entitled to rank as distinct species--or even, in many cases, as distinct genera. Yet we know, as a matter of fact, that all these differences have been produced by a process of artificial selection, or pairing, which has been continuously practised by horticulturists and breeders through a number of generations. It is the business of these men to note the individual organisms which show most variation in the directions required, and then to propagate from these individuals, in order that the progeny shall inherit the qualities desired. The results thus become cumulative from generation to generation, until we now have an astonishing manifestation of useful qualities on the one hand, and of beautiful qualities on the other, according as the organisms have been thus bred for purposes of use or for those of beauty.

Now it is immediately obvious that in these cases the process of artificial selection is precisely analogous to that of natural selection (and of sexual selection which will be considered later on), in all respects save one: the utility or the beauty which it is the aim of artificial selection continually to enhance, is utility or beauty in relation to the requirements or to the tastes of man; whereas the utility or the beauty which is produced by natural selection and sexual selection has reference only to the requirements or the tastes of the organisms themselves. But, with the exception of this one point of difference, the processes and the products are identical in kind. Persevering selection by man is thus proved to be capable of creating what are virtually new specific types, and this in any required direction. Hence, when we remember how severe is the struggle for existence in nature, it becomes impossible to doubt that selection by nature is able to do at least as much as artificial selection in the way of thus creating new types out of old ones. Artificial selection, indeed, notwithstanding the many and marvellous results which it has accomplished, can only be regarded as but a feeble imitation of natural selection, which must act with so much greater vigilance and through such immensely greater periods of time. In a word, the proved capabilities of artificial selection furnish, in its best conceivable form, what is called an argument a fortiori in favour of natural selection.

Or, to put it in another way, it may be said that for thousands of years mankind has been engaged in making a gigantic experiment to test, as it were by anticipation, the theory of natural selection. For, although this prolonged experiment has been carried on without any such intention on the part of the experimenters, it is none the less an experiment in the sense that its results now furnish an overwhelming verification of Mr. Darwin's theory. That is to say, they furnish overwhelming proof of the efficacy of the selective principle in the modification of organic types, when once this principle is brought steadily and continuously to bear upon a sufficiently long series of generations.

In order to furnish ocular evidence of the value of this line of verification, I have had the following series of drawings prepared. Another and equally striking series might be made of the products of artificial selection in the case of plants; but it seems to me that the case of animals is more than sufficient for the purpose just stated. Perhaps it is desirable to add that considerable care has been bestowed upon the execution of these portraits; and that in every case the latter have been taken from the most typical specimens of the artificial variety depicted. Those of them which have not been drawn directly from life are taken from the most authoritative sources; and, before being submitted to the engraver, they were all examined by the best judges in each department. In none of the groups, however, have I aimed at an exhaustive representation of all the varieties: I have merely introduced representatives of as many as the page would in each case accommodate.

The exigencies of space have prevented, in some of the groups, strict adherence to a uniform scale--with the result that contrasts between different breeds in respect of size are not adequately rendered. This remark applies especially to the dogs; for although the artist has endeavoured to draw them in perspective, unless the distance between those in the foreground and those in the background is understood to be more considerable than it appears, an inadequate idea is given of the relative differences of size. The most instructive of the groups, I think, is that of the Canaries; because the many and great changes in different directions must in this case have been produced by artificial selection in so comparatively short a time--the first mention of this bird that I can find being by Gesner, in the sixteenth century.

* * * * *

Now, it is surely unquestionable that in these typical proofs of the efficacy of artificial selection in the modification of specific types, we have the strongest conceivable testimony to the power of natural selection in the same direction. For it thus appears that wherever mankind has had occasion to operate by selection for a sufficiently long time--that is to say, on whatever species of plant or animal he chooses thus to operate for the purpose of modifying the type in any required direction,--the results are always more or less the same: he finds that all specific types lend themselves to continuous deflection in any particulars of structure, colour, &c., that he may desire to modify.

Nevertheless, to this parallel between the known effects of artificial selection, and the inferred effects of natural selection, two objections have been urged. The first is, that in the case of artificial selection the selecting agent is a voluntary intelligence, while in the case of natural selection the selecting agent is Nature herself; and whether or not there is any counterpart of man's voluntary intelligence in nature is a question with which Darwinism has nothing to do. Therefore, it is alleged, the analogy between natural selection and artificial selection fails ab initio, or at the fountain-head of the causes which are taken by the analogy to be respectively involved.

The second objection to the analogy is, that the products of artificial selection, closely as they may resemble natural species in all other respects, nevertheless present one conspicuous and highly important point of difference: they rarely, if ever, present the physiological character of mutual infertility, which is a character of extremely general occurrence in the case of natural species, even when these are most nearly allied.

I will deal with these two objections in the next chapter, where I shall be concerned with the meeting of all the objections which have ever been urged against the theory of natural selection. Meanwhile I am engaged only in presenting the general arguments which support the theory, and therefore mention these objections to one of them merely en passant. And I do so in order to pledge myself effectually to dispose of them later on, so that for the purposes of my present argument both these objections may be provisionally regarded as non-existent; which means, in other words, that we may provisionally regard the analogy between artificial selection and natural selection as everywhere logically intact.

* * * * *

To sum up, then, the results of the foregoing exposition thus far, what I hold to be the three principal, or most general, arguments in favour of the theory of natural selection, are as follows.

First, there is the a priori consideration that, if on independent grounds we believe in the theory of evolution at all, it becomes obvious that natural selection must have had some part in the process. For no one can deny the potent facts of heredity, variability, the struggle for existence, and survival of the fittest. But to admit these facts is to admit natural selection as a principle which must be, at any rate, one of the factors of organic evolution, supposing such evolution to have taken place. Next, when we turn from these a priori considerations, which thus show that natural selection must have been concerned to some extent in the process of evolution, we find in organic nature evidence a posteriori of the extent to which this principle has been thus concerned. For we find that among all the countless millions of adaptive structures which are to be met with in organic nature, it is an invariable rule that they exist in relation to the needs of the particular species which present them: they never have any primary reference to the needs of other species. And as this extraordinarily large and general fact is exactly what the theory of natural selection would expect, the theory is verified by the fact in an extraordinarily cogent manner. In other words, the fact goes to prove that in all cases where adaptive structures or instincts are concerned, natural selection must have been either the sole cause at work, or, at the least, an influence controlling the operation of all other causes.

Lastly, an actually experimental verification of the theory has been furnished on a gigantic scale by the operations of breeders, fanciers, and horticulturists. For these men, by their process of selective accumulation, have empirically proved what immense changes of type may thus be brought about; and so have verified by anticipation, and in a most striking manner, the theory of natural selection--which, as now so fully explained, is nothing more than a theory of cumulative modifications by means of selective breeding.

So much, then, by way of generalities. But perhaps the proof of natural selection as an agency of the first importance in the transmutation of species may be best brought home to us by considering a few of its applications in detail. I will therefore devote the rest of the present chapter to considering a few cases of this kind.

There are so many large fields from which such special illustrations may be supplied, that it is difficult to decide which of them to draw upon. For instance, the innumerable, always interesting, and often astonishing adaptations on the part of flowers to the fertilising agency of insects, has alone given rise to an extensive literature since the time when Darwin himself was led to investigate the subject by the guidance of his own theory. The same may be said of the structures and movements of climbing plants, and in short, of all the other departments of natural history where the theory of natural selection has led to the study of the phenomena of adaptation. For in all these cases the theory of natural selection, which first led to their discovery, still remains the only scientific theory by which they can be explained. But among all the possible fields from which evidences of this kind may be drawn, I think the best is that which may be generically termed defensive colouring. To this field, therefore, I will restrict myself. But, even so, the cases to be mentioned are but mere samples taken from different divisions of this field; and therefore it must be understood at the outset that they could easily be multiplied a hundred-fold.

Protective Colouring.

A vast number of animals are rendered more or less inconspicuous by resembling the colours of the surfaces on which they habitually rest. Such, for example, are grouse, partridges, rabbits, &c. Moreover, there are many cases in which, if the needs of the creature be such that it must habitually frequent surfaces of different colours, it has acquired the power of changing its colour accordingly--e. g. cuttle-fish, flat-fish, frogs, chameleons, &c. The physiological mechanism whereby these adaptive changes of colour are produced differs in different animals; but it is needless for our purposes to go into this part of the subject. Again, there are yet other cases where protective colouring which is admirably suited to conceal an animal through one part of the year, would become highly conspicuous during another part of it--namely, when the ground is covered with snow. Accordingly, in these cases the animals change their colour in the winter months to a snowy white: witness stoats, mountain hares, ptarmigan, &c. (Fig. 108.)

Now, it is sufficiently obvious that in all these classes of cases the concealment from enemies or prey which is thus secured is of advantage to the animals concerned; and, therefore, that in the theory of natural selection we have a satisfactory theory whereby to explain it. And this cannot be said of any other theory of adaptive mechanisms in nature that has ever been propounded. The so-called Lamarckian theory, for instance, cannot be brought to bear upon the facts at all; and on the theory of special creation it is unintelligible why the phenomena of protective colouring should be of such general occurrence. For, in as far as protective colouring is of advantage to the species which present it, it is of corresponding disadvantage to those other species against the predatory nature of which it acts as a defence. And, of course, the same applies to yet other species, if they serve as prey. Moreover, the more minutely this subject is investigated in all its details, the more exactly is it found to harmonise with the naturalistic interpretation[38].

[38] Were it not that some of Darwin's critics have overlooked the very point wherein the great value of protective colouring as evidence of natural selection consists, it would be needless to observe that it does so in the minuteness of the protective resemblance which in so many cases is presented. Of course where the resemblance is only very general, the phenomena might be ascribed to mere coincidence, of which the instincts of the animal have taken advantage. But in the measure that the resemblance becomes minutely detailed, the supposition of mere coincidence is excluded, and the agency of some specially adaptive cause demonstrated. Again, it is almost needless to say, no real difficulty is presented (as has been alleged) by the cases above quoted of seasonal imitations, on the ground that natural selection could not act alternately on the same individual. Natural selection is

not supposed to act alternately on the same individual. It is supposed to act always in the same manner, and if, as in the case of a regularly recurring change in the colours of the environment, correspondingly recurrent changes are required to appear in the colours of the animals, natural selection sets its premium upon those individuals the constitutions of which best lend themselves to seasonal changes of the needful kind--probably under the influence of stimuli supplied by the changes of external conditions (temperature, moisture, &c.).

In the first place, we always find a complete correspondence between imitative colouring and instinctive endowment. If a caterpillar exactly resembles the colour of a twig, it also presents the instinct of habitually reposing in the attitude which makes it most resemble a twig--standing out from the branch on which it rests at the same angle as is presented by the real twigs of the tree on which it lives.

Here, again, is a bird protectively coloured so as to resemble stones upon the rough ground where it habitually lives; and the drawing shows the attitude in which the bird instinctively reposes, so as still further to increase its resemblance to a stone. (Fig. 109.)

To take only one other instance, hares and rabbits, like grouse and partridges--or like the plover just alluded to,--instinctively crouch upon those surfaces the colours of which they resemble; and I have often remarked that if, on account of any individual peculiarity of coloration, the animal is not able thus to secure concealment, it nevertheless exhibits the instinct of crouching which is of benefit to all its kind, although, from the accident of its own abnormal colouring, this instinct is then actually detrimental to the animal itself. For example, every sportsman must have noticed that the somewhat rare melanic variety of the common rabbit will crouch as steadily as the normal brownish-gray type, notwithstanding that, owing to its abnormal colour, a "nigger-rabbit" thus renders itself the most conspicuous object in the landscape. In all such cases, of course, there has been a deviation from the normal type in respect of colour, with the result that the inherited instinct is no longer in tune with the other endowments of the animal. Such a variation of colour, therefore, will tend to be suppressed by natural selection; while any variations which may bring the animal still more closely to resemble its habitual surroundings will be preserved. Thus we can understand the truly wonderful extent to which this principle of protective colouring has been carried in many cases where the need of it has been most urgent.

Not only colour, but structure, may be profoundly modified for the purposes of protective concealment. Thus, caterpillars which resemble twigs do so not only in respect of colour, but also of shape; and this even down to the most minute details in cases where the adaptation is most complete: certain butterflies and leaf-insects so precisely resemble the leaves upon which, or among which, they live, that it is almost impossible to detect them in the foliage--not only the colour, the shape, and the venation being all exactly imitated, but in some cases even the defects to which the leaves are liable, in the way of fungoid growths, &c. There are other insects which with similar exactness resemble moss, lichens, and so forth. A species of fish secures a complete resemblance to bunches of sea-weed by a frond-like modification of all its appendages, and so on through many other instances. Now, in all such cases where there is so precise an imitation, both in colour and structure, it seems impossible to suggest any other explanation of the facts than the one which is supplied by Mr. Darwin's theory--namely, that the more perfect the resemblance is caused to become through the continuous influence of natural selection always picking out the best imitations, the more highly discriminative becomes the perception of those enemies against the depredations of which this peculiar kind of protection is developed; so that, in virtue of this action and re-action, eventually we have a degree of imitation which renders it almost impossible for a naturalist to detect the animal when living in its natural environment.

In strange and glaring contrast to all these cases of protective colouring, stand other cases of conspicuous colouring. Thus, for example, although there are numberless species of caterpillars which present in an astonishing degree the phenomena of protective colouring, there are numberless other species which not only fail to present these phenomena in any degree, but actually go to the opposite extreme of presenting colours which appear to have been developed for the sake of their conspicuousness. At all events, these caterpillars are usually the most conspicuous objects in their surroundings, and therefore in the early days of Darwinism they were regarded by Darwin himself as presenting a formidable difficulty in the way of his theory. To Mr. Wallace belongs the merit of having cleared up this difficulty in an extraordinarily successful manner. He virtually reasoned thus.

If the raison d'être of protective colouring be that of concealing agreeably flavoured caterpillars from the eye-sight of birds, may not the raison d'être of conspicuous colouring be that of protecting disagreeably flavoured caterpillars from any possibility of being mistaken by birds? Should this be the case, of course the more conspicuous the colouring the better would it be for the caterpillars presenting it. Now as soon as this suggestion was acted upon experimentally, it was found to be borne out by facts. Birds could not be induced to eat caterpillars of the kinds in question; and there is now no longer any doubt that their conspicuous colouring is correlated with their distastefulness to birds, in the same way as the inconspicuous or imitative colouring of other caterpillars is correlated with their tastefulness to birds. Here then is yet another instance, added to those already given, of the verification yielded to the theory of natural selection by its proved competency as a guide to facts in nature; for assuredly this particular class of facts would never have been suspected but for its suggestive agency.

As in the case of protective imitation, so in this case of warning conspicuousness, not only colour, but structure may be greatly modified for the purpose of securing immunity from attack. Here, of course, the object is to assume, as far as possible, a touch-me-not appearance; so that, although destitute of any real means of offence, the creatures in question present a fictitiously dangerous aspect. As the Devil's-coach-horse turns up his stingless tail when threatened by an enemy, so in numberless ways do many harmless animals of all classes pretend to be formidable. But the point now is that these instincts of self-defence are often helped out by structural modifications, expressly and exclusively adapted to this end. For example, what a remarkable series of protective adjustments occurs in the life-history of the Puss Moth--culminating with so comical an instance of the particular device now under consideration as the following. I quote the facts from Mr. E. B. Poulton's admirable book on The Colours of Animals (pp. 269-271).

The larva of the Puss Moth (Cerura vinula) is very common upon poplar and willow. The circular dome-like eggs are laid, either singly or in little groups of two or three, upon the upper side of the leaf, and being of a reddish colour strongly suggest the appearance of little galls, or the results of some other injury to the leaf. The youngest larv?are black, and also rest upon the upper surface of the leaf, resembling the dark patches which are commonly seen in this position. As the larva grows, the apparent black patch would cover too large a space, and would lead to detection if it still occupied the whole surface of the body. The latter gains a green ground-colour which harmonises with the leaf, while the dark marking is chiefly confined to the back. As growth proceeds the relative amount of green increases, and the dark mark is thus prevented from attaining a size which would render it too conspicuous. In the last stage of growth the green larva becomes very large, and usually rests on the twigs of its food-plant (Fig. 111). The dark colour is still present on the back but is softened to a purplish tint, which tends to be replaced by a combination of white and green in many of the largest larv? Such a larva is well concealed by General Protective Resemblance, and one may search a long time before finding it, although assured of its presence from the stripped branches of the food-plant and the f 鎰 es on the ground beneath.

As soon as a large larva is discovered and disturbed it withdraws its head into the first body-ring, inflating the margin, which is of a bright red colour. There are two intensely black spots on this margin in the appropriate position for eyes, and the whole appearance is that of a large flat face extending to the outer edge of the red margin (see Fig. 112). The effect is an intensely exaggerated caricature of a vertebrate face, which is probably alarming to the vertebrate enemies of the caterpillar. The terrifying effect is therefore mimetic. The movements entirely depend on tactile impressions: when touched ever so lightly a healthy larva immediately assumes the terrifying attitude, and turns so as to present its full face towards the enemy; if touched on the other side or on the back it instantly turns its face in the appropriate direction. The effect is also greatly strengthened by two pink whips which are swiftly protruded from the prongs of the fork in which the body terminates. The prongs represent the last pair of larval legs which have been greatly modified from their ordinary shape and use. The end of the body is at the same time curved forward over the back (generally much further than in Fig. 112), so that the pink filaments are brandished above the head.

Mimicry.

Lastly, these facts as to imitative and conspicuous colouring lead on to the yet more remarkable facts of what is called mimicry. By mimicry is meant the imitation in form and colour of one species by another, in order that the imitating species may be mistaken for the imitated, and thus participate in some advantage which the latter enjoys. For instance, if, as in the case of the conspicuously-coloured caterpillars, it is of advantage to an ill-savoured species that it should hold out a warning to enemies, clearly it may be of no less advantage to a well-savoured species that it should borrow this flag, and thus be mistaken for its ill-savoured neighbour. Now, the extent to which this device of mimicry is carried is highly remarkable, not only in respect of the number of its cases, but also in respect of the astonishing accuracy which in most of these cases is exhibited by the imitation. There need be little or virtually no zoological affinity between the imitating and the imitated forms; that is to say, in some cases the zoological affinity is not closer than ordinal, and therefore cannot possibly be ascribed to kinship. Like all the other branches of the general subject of protective resemblance in form or colouring, this branch has already been so largely illustrated by previous writers, that, as in the previous cases, I need only give one or two examples. Those which I choose are chosen on account of the colours concerned not being highly varied or brilliant, and therefore lending themselves to less ineffectual treatment by wood-engraving than is the case where attempts are made to render by this means even more remarkable instances.

It is surely apparent, without further comment, that it is impossible to imagine stronger evidence in favour of natural selection as a true cause in nature, than is furnished by this culminating fact in the matter of protective resemblance, whereby it is shown that a species of one genus, family, or even order, will accurately mimic the appearance of a species belonging to another genus, family, or order, so as to deceive its natural enemies into mistaking it for a creature of so totally different a kind. And it must be added that while this fact of mimicry is of extraordinarily frequent occurrence, there can be no possibility of our mistaking its purpose. For the fact is never observable except in the case of species which occupy the same area or district.

Such being what appears to me the only reasonable view of the matter, I will now conclude this chapter on the evidences of natural selection as at all events the main factor of organic evolution, by simply adding illustrations of two further cases of mimicry, which are perhaps even more remarkable than any of the foregoing examples. The first of the two (Fig. 115) speaks for itself. The second will be rendered intelligible by the following few words of explanation.

There are certain ants of the Amazons which present the curious instinct of cutting off leaves from trees, and carrying them like banners over their heads to the hive, as represented in Fig. 116, B, where one ant is shown without a leaf, and the others each with a leaf. Their object in thus collecting leaves is probably that of growing a fungus upon the "soil" which is furnished by the leaves when decomposing. But, be this as it may[39], the only point we are now concerned with is the appearance which these ants present when engaged in their habitual operation of carrying leaves. For it has been recently observed by Mr. W. L. Sclater, that in the localities where these hymenopterous insects occur, there occurs also a homopterous insect which mimics the ant, leaf and all, in a wonderfully deceptive manner. The leaf is imitated by the thin flattened body of the insect, "which in its dorsal aspect is so compressed laterally that it is no thicker than a leaf, and terminates in a sharp jagged edge." The colour is exactly the same as that of a leaf, and the brown legs show themselves beneath the green body in just the same way as those of the ant show themselves beneath the leaf. So that both the form and the colouring of the homopterous insect has been brought to resemble, with singular exactness, those belonging to a different order of insect, when the latter is engaged in its peculiar avocation. A glance at the figure is enough to show the means employed and the result attained. In A, an ant and its mimic are represented as about 2-1/2 times their natural size, and both proceeding in the same direction. It ought to be mentioned, however, that in reality the margin of the leaf is seldom allowed to retain its natural serrations as here depicted: the ants usually gnaw the edge of the real leaf, so that the margin of the false one bears an even closer resemblance to it than the illustration represents. B is a drawing from life of a group of five ants carrying leaves, and their mimic walking beside them[40].

[39] For a full account of this instinct and its probable purpose, see Animal Intelligence, pp. 93-6.

[40] Both drawings are reproduced from Mr. Poulton's paper upon the subject (Proc. Zool. Soc., June 16, 1891).

CHAPTER IX.

CRITICISMS OF THE THEORY OF NATURAL SELECTION.

I will now proceed to consider the various objections and difficulties which have hitherto been advanced against the theory of natural selection.

Very early in the day Owen hurled the weight of his authority against the new theory, and this with a strength of onslaught which was only equalled by its want of judgment. Indeed, it is painfully apparent that he failed to apprehend the fundamental principles of the Darwinian theory. For he says:--

Natural Selection is an explanation of the process [of transmutation] of the same kind and value as that which has been proffered of the mystery of "secretion." For example, a particular mass of matter in a living animal takes certain elements out of the blood, and rejects them as "bile." Attributes were given to the liver which can only be predicated of the whole animal; the "appetency" of the liver, it was said, was for the elements of bile, and "biliosity," or the "hepatic sensation," guided the gland to their secretion. Such figurative language, I need not say, explains absolutely nothing of the nature of bilification[41].

[41] Anatomy of Vertebrates, vol. iii. p. 794.

Assuredly, it was needless for Owen to say that figurative language of this kind explains nothing; but it was little less than puerile in him to see no more in the theory of natural selection than such a mere figure of speech. To say that the liver selects the elements of bile, or that nature selects specific types, may both be equally unmeaning re-statements of facts; but when it is explained that the term natural selection, unlike that of "hepatic sensation," is used as a shorthand expression for a whole group of well-known natural causes--struggle, variation, survival, heredity,--then it becomes evidence of an almost childish want of thought to affirm that the expression is figurative and nothing more. The doctrine of natural selection may be a huge mistake; but, if so, this is not because it consists of any unmeaning metaphor: it can only be because the combination of natural causes which it suggests is not of the same adequacy in fact as it is taken to be in theory.

Owen further objected that the struggle for existence could only act as a cause of the extinction of species, not of their origination--a view of the case which again shows on his part a complete failure to grasp the conception of Darwinism. Acting alone, the struggle for existence could only cause extermination; but acting together with variation, survival, and heredity, it may very well--for anything that Owen, or others who followed in this line of criticism, show to the contrary--have produced every species of plant and animal that has ever appeared upon the face of the earth.

Another and closely allied objection is, that the theory of natural selection "personifies an abstraction." Or, as the Duke of Argyll states it, the theory is "essentially the image of mechanical necessity concealed under the clothes, and parading in the mask, of mental purpose. The word 'natural' suggests Matter, and the physical forces. The word 'selection' suggests Mind, and the powers of choice." This, however, is a mere quarrelling about words. Darwin called the principle which he had discovered by the name natural selection in order to mark the analogy between it and artificial selection. No doubt in this analogy there is not necessarily supposed to be in nature any counterpart to the mind of the breeder, nor, therefore, to his powers of intelligent choice. But there is no need to limit the term selection (se and lego, Gr. [Greek: leg 鬭]) to powers of intelligent choice. As previously remarked, a bank of sea-weed on the sea-shore may be said to have been selected by the waves from all the surrounding sand and stones. Similarly, we may say that grain is selected from chaff by the wind in the process of winnowing corn. Or, if it be thought that there is any ambiguity involved in such a use of the term in the case of "Natural Selection," there is no objection to employing the phrase which has been coined by Mr. Spencer as its equivalent--namely, "Survival of the Fittest." The point of the theory is, that those organisms which are best suited to their surroundings are allowed to live and to propagate, while those which are less suited are eliminated; and whether we call this process a process of selection, or call it by any other name, is clearly immaterial.

A material question is raised only when it is asked whether the process is one that can be ascribed to causation strictly natural. It is often denied that such is the case, on the ground that natural selection does not originate the variations which it favours, but depends upon the variations being supplied by some other means. For, it is said, all that natural selection does is to preserve the suitable variations after they have arisen. Natural selection does not cause these suitable variations; and therefore, it is argued, Darwin and his followers are profoundly mistaken in representing the principle as one which produces adaptations. Now, although this objection has been put forward by some of the most intelligent minds in our generation, it appears to me to betoken some extraordinary failure to appreciate the very essence of

Darwinian doctrine. No doubt it is perfectly true that natural selection does not produce variations of any kind, whether beneficial or otherwise. But if it be granted that variations of many kinds are occurring in every generation, and that natural selection is competent to preserve the more favourable among them, then it appears to me unquestionable that this principle of selection deserves to be regarded as, in the full sense of the word, a natural cause. The variations being expressly regarded by the theory as more or less promiscuous[42], survival of the fittest becomes the winnowing fan, whose function it is to eliminate all the less fit in each generation, in order to preserve the good grain, out of which to constitute the next generation. And as this process is supposed to be continuous through successive generations, its action is supposed to be cumulative, till from the eye of a worm there is gradually developed the eye of an eagle. Therefore it follows from these suppositions (which are not disputed by the present objection), that if it had not been for the process of selection, such development would never have been begun; and that in the exact measure of its efficiency will the development proceed. But any agency without the operation of which a result cannot take place may properly be designated the cause of that result: it is the agency which, in co-operation with all the other agencies in the cosmos, produces that result.

[42] The degree in which variability is indefinite, or, on the contrary, determinate, is a question which is not yet ripe for decision--nor even, in my opinion, for discussion. But I may here state the following general principles with regard to it.

(1) It is evident that up to some point or another variations must be pre-determined in definite lines. Men do not gather grapes from thorns, figs from thistles, nor even moss-roses from sweet-briars. In other words, "the nature of the organism" in all cases necessitates the limiting of variations within certain bounds.

(2) But when the question is as to what these bounds may be, we can only answer in a general way that, according to the general theory of evolution, they must be such as are imposed by heredity, coupled with the degree to which external conditions of life (and possibly also use-inheritance) are capable, in given cases, of modifying congenital characters. These are the only causes which the theory of descent can consistently recognise as

producing variations in determinate directions.

(3) Inasmuch as variation presupposes the existence of parts that vary, and inasmuch as the variation of parts can only be in the alternative directions of increase or decrease around an average, it follows that, in the first instance at all events, every variation, if determinate, must be so only in one or other of these two opposite directions.

(4) In as far as variations are summated in successive generations, so as eventually to give rise to new structures, organs, mechanisms, &c., natural selection is theoretically competent to explain the facts, without our having to postulate the operation of unknown causes producing variations in determinate lines,--or not further than is stated in paragraphs 1 and 2.

(5) Nevertheless, it does not follow that there are not such other unknown causes; and, if there are, of course the importance of natural selection as a cause of adaptive modification would be limited in proportion to their number and the extent of their operation. But it is for those who, like the late Professors Asa Gray and N 鋊 eli, maintain the existence of such causes, to substantiate their belief by indicating them.

Take any analogous case. The selective agency of specific gravity which is utilised in gold-washing does not create the original differences between gold-dust and dust of all other kinds. But these differences being presented by as many different bodies in nature, the gold-washer takes advantage of the selective agency in question, and, by using it as a cause of segregation, is enabled to separate the gold from all the earths with which it may happen to be mixed. So far as the objects of the gold-washer are concerned, it is immaterial with what other earths the gold-dust may happen to be mixed. For although gold-dust may occur in intimate association with earths of various kinds in various proportions, and although in each case the particular admixture which occurs must have been due to definite causes, these things, in relation to the selective process of the washer, are what is called accidental: that is to say, they have nothing to do with the causative action of the selective process. Now, in precisely the same sense Darwin calls the multitudinous variations of plants and animals accidental. By so calling them he expressly says he does not suppose them to be accidental in the sense of not all being due to definite causes. But they are accidental in relation to the sifting process of natural selection: all that they have to do is to furnish the promiscuous material on which this sifting process acts.

Or let us take an even closer analogy. The power of selective breeding by man is so wonderful, that in the course of successive generations all kinds of peculiarities as to size, shape, colour, special appendages or abortions, &c., can be produced at pleasure, as we saw in the last chapter. Now all the promiscuous variations which are supplied to the breeder, and out of which, by selecting only those that are suited to his purpose, he is able to produce the required result--all those promiscuous variations, in relation to that purpose, are accidental. Therefore the selective agency of the breeder deserves to be regarded as the cause of that which it produces, or of that which could not have been produced but for the operation of such agency. But where is the difference between artificial and natural selection in this respect? And, if there is no difference, is not natural selection as much entitled to be regarded as a true cause of the origin of natural species, as artificial selection is to be regarded as a true cause of our domesticated races? Here, as in the case of the previous illustration, if there be any ambiguity in speaking of variations as accidental, it arises from the incorrect or undefined manner in which the term "accidental" is used by Darwin's critics. In its original and philosophically-correct usage, the term "accident" signifies a property or quality not essential to our conception of a substance: hence, it has come to mean anything that happens as a result of unforeseen causes--or, lastly, that which is causeless. But, as we know that nothing can happen without causes of some kind, the term "accident" is divested of real meaning when it is used in the last of these senses. Yet this is the sense that is sought to be placed upon it by the objection which we are considering. If the objectors will but understand the term in its correct philosophical sense--or in the only sense in which it presents any meaning at all,--they will see that Darwinians are both logically and historically justified in employing the word "accidental" as the word which serves most properly to convey the meaning that they intend--namely, variations due to causes accidental to the struggle for existence. Similarly, when it is said that variations are "spontaneous," or even "fortuitous," nothing further is meant than that we do not know the causes which lead to them, and that, so far as the principle of selection is concerned, it is immaterial what these causes may be. Or, to revert to our former illustration, the various weights of different kinds of earths are no doubt all due to definite causes; but, in relation to the selective action of the gold-washer, all the different weights of whatever kinds of earth he may happen to include in his washing-apparatus are, strictly speaking, accidental. And as at different washings he meets with different proportions of heavy earths with light ones, and as these "variations" are immaterial to him, he may colloquially speak of them as "fortuitous," or due to "chance," even though he knows that at each washing they must have been determined by definite causes.

More adequately to deal with this merely formal objection, however, would involve more logic-chopping than is desirable on the present occasion. But I have already dealt with it fully elsewhere,--viz. in The Contemporary Review for June, 1888, to which therefore I may refer any one who is interested in dialectics of this kind[43].

[43] Within the last few months this objection has been presented anew by Mr. D. Syme, whose book On the Modification of Organisms exhibits a curious combination of shrewd criticisms with almost ludicrous misunderstandings. One of the latter it is necessary to state, because it pervades the quotation which I am about to supply. He everywhere compares "natural selection" with "the struggle for existence," uses them as convertible terms, and while absurdly stating that "Darwin defines natural selection as the struggle for existence," complains of "the liability of error, both on his own part and on the part of his readers," which arises from his not having everywhere adhered to this definition! (p. 8).

"Darwin has put forth two distinct and contradictory theories of the functions of natural selection. According to the one theory natural selection is selective or preservative, and nothing more. According to the other theory natural selection creates the variations(!) ... It certainly seems absurd to speak of natural selection, or the struggle for existence, as selective or preservative, for the struggle for existence does not preserve at all, not even the fit variations, as both the fit and the unfit struggle for existence, the unfit naturally more than the fit, and the fit are preserved, not in consequence of the struggle, but in consequence of their fitness. Suppose two varieties of the same species are driven, by an increase of their numbers, to seek for subsistence in a colder region than they have been accustomed to, and that one of these varieties had a hardier constitution than the other; and let us suppose that the former withstood the severe climate better than the latter, and consequently survived, while the other perished. In this case the hardier survived, not because of the struggle, but because it had a constitution better adapted to the climate. I wish to ascertain if a certain metal in my possession is gold or some baser metal, and I apply the usual test; but the mere fact of my testing this metal would not make it gold or any other kind of metal."

I have thought it worth while to quote this passage for the sake of showing the extraordinary confusion of mind which still prevails on the part of Darwin's critics, even with reference to the very fundamental parts of his theory. For, as I have said, the writer of this passage shows himself a shrewd critic in some other parts of his essay, where he is not engaged especially on the theory of natural selection.

I will now pass on to consider another misconception of the Darwinian theory, which is very prevalent in the public mind. It is virtually asked, If some species are supposed to have been improved by natural selection, why have not all species been similarly improved? Why should not all invertebrated animals have risen into vertebrated? Or why should not all monkeys have become men?

The answers are manifold. In the first place, it by no means follows that because an advance in organization has proved itself of benefit in the case of one form of life, therefore any or every other form would have been similarly benefited by a similar advance. The business of natural selection is to bring this and that form of life into the closest harmony with its environment that all the conditions of the case permit. Sometimes it will happen that the harmony will admit of being improved by an improvement of organization. But just as often it will happen that it will be best secured by leaving matters as they are. If, therefore, an organism has already been brought into a tolerably full degree of harmony with its environment, natural selection will not try to change it so long as the environment remains unchanged; and this, no doubt, is the reason why some species have survived through enormous periods of geological time without having undergone any change. Again, as we saw in a previous chapter, there are yet other cases where, on account of some change in the environment or even in the habits of the organisms themselves, adaption will be best secured by an active reversal of natural selection, with the result of causing degeneration.

But, it is sometimes further urged, there are cases where we cannot doubt that improvement of organization would have been of benefit to species; and yet such improvement has not taken place--as, for instance, in the case all monkeys not turning into men. Here, however, we must remember that the operation of natural selection in any case depends upon a variety of highly complex conditions; and, therefore, that the fact of all those conditions having been satisfied in one instance is no reason for concluding that they must also have been satisfied in other instances. Take, for example, the case of monkeys passing into men. The wonder to me appears to be that this improvement should have taken place in even one line of descent; not that, having taken place in one line, it should not also have taken place in other lines. For how enormously complex must have been the conditions--physical, anatomical, physiological, psychological, sociological--which by their happy conjunction first began to raise the inarticulate cries of an ape into the rational speech of a man. Therefore, the more that we appreciate the superiority of a man to an ape, the less ought we to countenance this supposed objection to Darwin's theory--namely, that natural selection has not effected the change in more than one line of descent.

Even in the case of two races of mankind where one has risen higher in the scale of civilization than another, it is now generally impossible to assign the particular causes of the difference; much more, then, must this be impossible in the case of still more remote conditions which have led to the divergence of species. The requisite variations may not have arisen in the one line of descent which did arise in the other; or if they did arise in both, some counterbalancing disadvantages may have attended their initial development in the one case which did not obtain in the other. In short, where so exceedingly complex a play of conditions are concerned, the only wonder would be if two different lines of descent had happened to present two independent and yet perfectly parallel lines of history.

These general considerations would apply equally to the great majority of other cases where some types have made great advances upon others, notwithstanding that we can see no reason why the latter should not in this respect have imitated the former. But there is yet a further consideration which must be taken into account. The struggle for existence is always most keen between closely allied species, because, from the similarity of their forms, habits, needs, &c., they are in closest competition. Therefore it often happens that the mere fact of one species having made an advance upon others of itself precludes the others from making any similar advance: the field, so to speak, has already been occupied as regards that particular improvement, and where the struggle for existence is concerned possession is emphatically nine points of the law. For example, to return to the case of apes becoming men, the fact of one rational species having been already evolved (even if the rational faculty were at first but dimly nascent) must make an enormous change in the conditions as regards the possibility of any other such species being subsequently evolved--unless, of course, it be by way of descent from the rational one. Or, as Sir Charles Lyell has well put it, two rational species can never coexist on the globe, although the descendants of one rational species may in time become transformed into another single rational species[44].

[44] Principles of Geology, vol. ii. p. 487 (11th ed.).

In view of such considerations, another and exactly opposite objection has sometimes been urged--viz. that we ought never to find inferior forms of organization in company with superior, because in the struggle for existence the latter ought to have exterminated the former. Or, to quote the most recent expression of this view, "in every locality there would only be one species, and that the most highly organized; and thus a few superior races would partition the earth amongst them to the entire exclusion of the innumerable varieties, species, genera, and orders which now inhabit it[45]." Of course to this statement it would be sufficient to enquire, On what would these few supremely organized species subsist? Unless manna fell from heaven for their especial benefit, it would appear that such forms could under no circumstances be the most improved forms; in exterminating others on such a scale as this, they would themselves be quickly, and very literally, improved off the face of the earth. But even when the statement is not made in so extravagant a form as this, it must necessarily be futile as an objection unless it has first been shown that we know exactly all the conditions of the complex struggle for existence between the higher and lower forms in question. And this it is impossible that we ever can know. The mere fact that one form has been changed in virtue of this struggle must in many cases of itself determine a change in the conditions of the struggle. Again, the other and closely allied forms (and these furnish the best grounds for the objection) may also have undergone defensive changes, although these may be less conspicuous to our observation, or perhaps less suggestive of "improvement" to our imperfect means of judging. Lastly, not to continue citing an endless number of such considerations, there is the broad fact that it is only to those cases where, for some reason or another, the lower forms have not been exposed to a struggle of fatal intensity, that the objection applies. But we know that in millions of other cases the lower (i. e. less fitted) forms have succumbed, and therefore I do not see that the objection has any ground to stand upon. That there is a general tendency for lower forms to yield their places to higher is shown by the gradual advance of organization throughout geological time; for if all the inferior forms had survived, the earth could not have contained them, unless she had been continually growing into something like the size of Jupiter. And if it be asked why any of the inferior forms have survived, the answer has already been given, as above.

[45] Syme, on the Modification of Organisms, p. 46.

There is only one other remark to be made in this connexion. Mr. Syme chooses two cases as illustrations of the supposed difficulty. These are sufficiently diverse--viz. Foraminifera and Man. Touching the former, there is nothing that need be added to the general answer just given. But with regard to the latter it must be observed that the dominion of natural selection as between different races of mankind is greatly restricted by the presence of rationality. Competition in the human species is more concerned with wits and ideas than with nails and teeth; and therefore the "struggle" between man and man is not so much for actual being, as for well-being. Consequently, in regard to the present objection, the human species furnishes the worst example that could have been chosen.

* * * * *

Hitherto I have been considering objections which arise from misapprehensions of Darwin's theory. I will now go on to consider a logically sound objection, which nevertheless is equally futile, because, although it does not depend on any misapprehension of the theory, it is not itself supported by fact.

The objection is the same as that which we have already considered in relation to the general theory of descent--namely, that similar organs or structures are to be met with in widely different branches of the tree of life. Now this would be an objection fatal to the theory of natural selection, supposing these organs or structures in the cases compared are not merely analogous, but also homologous. For it would be incredible that in two totally different lines of descent one and the same structure should have been built up independently by two parallel series of variations, and that in these two lines of descent it should always and independently have ministered to the same function. On the other hand, there would be nothing against the theory of natural selection in the fact that two structures, not homologous, should come by independent variation in two different lines of descent to be adapted to perform the same function. For it belongs to the very essence of the theory of natural selection that a useful function should be secured by favourable variations of whatever structural material may happen to be presented by different organic types. Flying, for instance, is a very useful function, and it has been developed independently in at least four different lines of descent--namely, the insects, reptiles, birds, and mammals. Now if in all, or indeed in any, of these four cases the wings had been developed on the same anatomical pattern, so as not only to present the analogical resemblance which it is necessary that they should present in order to discharge their common function of flying, but likewise an homologous or structural resemblance, showing that they had been formed on the same anatomical "plan,"--if such has been the case, I say, the theory of natural selection would certainly be destroyed.

Now it has been alleged by competent naturalists that there are several such cases in organic nature. We have already noticed in a previous chapter (pp. 58, 59), that Mr. Mivart has instanced the eye of the cuttle-fish as not only analogous to, but also homologous with, the eye of a true fish--that is to say, the eye of a mollusk with the eye of a vertebrate. And he has also instanced the remarkable resemblance of a shrew to a mouse--that is, of an insectivorous mammal to a rodent--not to mention other cases. In the chapter alluded to these instances of homology, alleged to occur in different branches of the tree of life, were considered with reference to the process of organic evolution as a fact: they are now being considered with reference to the agency of natural selection as a method. And just as in the former case it was shown, that if any such alleged instances could be proved, the proof would be fatal to the general theory of organic evolution by physical causes, so in the present case, if this could be proved, it would be equally fatal to the more special theory of natural selection. But, as we have before seen, no single case of this kind has ever been made out; and, therefore, not only does this supposed objection fall to the ground, but in so doing it furnishes an additional argument in favour of natural selection. For in the earlier chapter just alluded to I showed that this great and general fact of our nowhere being able to find two homologous structures in different branches of the tree of life, was the strongest possible testimony in favour of the theory of evolution. And, by parity of reasoning, I now adduce it as equally strong evidence of natural selection having been the cause of adaptive structures, independently developed in all the different lines of descent. For the alternative is between adaptations having been caused by natural selection or by supernatural design. Now, if adaptations were caused by natural selection, we can very well understand why they should never be homologous in different lines of descent, even in cases where they have been brought to be so closely analogous as to have deceived so good a naturalist as Mr. Mivart. Indeed, as I have already observed, so well can we understand this, that any single instance to the contrary would be sufficient to destroy the theory of natural selection in toto, unless the structure be one of a very simple type. But on the other hand, it is impossible to suggest any rational explanation why, if all adaptations are due to supernatural design, such scrupulous care should have been taken never to allow homologous adaptations to occur in different divisions of the animal or vegetable kingdoms. Why, for instance, should the eye of a cuttle-fish not have been constructed on the same ideal pattern as that of vertebrate? Or why, among the thousands of vertebrated species, should no one of their eyes be constructed on the ideal pattern that was devised for the cuttle-fish? Of course it may be answered that perhaps there was some hidden reason why the design should never have allowed an adaptation which it had devised for one division of organic nature to appear in another--even in cases where the new design necessitated the closest possible resemblance in everything else, save in the matter of anatomical homology. Undoubtedly such may have been the case--or rather such must have been the case--if the theory of special design is true. But where the question is as to the truth of this theory, I think there can be no doubt that its rival gains an enormous advantage by being able to explain why the facts are such as they are instead of being obliged to take refuge in hypothetical possibilities of a confessedly unsubstantiated and apparently unsubstantial kind.

Therefore, as far as this objection to the theory of natural selection is concerned--or the allegation that homologous structures occur in different divisions of organic nature--not only does it fall to the ground, but positively becomes itself converted into one of the strongest arguments in favour of the theory. As soon as the allegation is found to be baseless, the very fact that it cannot be brought to bear upon any one of all the millions of adaptive structures in organic nature becomes a fact of vast significance on the opposite side.

* * * * *

The next difficulty to which I shall allude is that of explaining by the theory of natural selection the preservation of the first beginnings of structures which are then useless, though afterwards, when more fully developed, they become useful. For it belongs to the very essence of the theory of natural selection, that a structure must be supposed already useful before it can come under the influence of natural selection: therefore the theory seems incapable of explaining the origin and conservation of incipient organs, or organs which are not yet sufficiently developed to be of any service to the organisms presenting them.

This objection is one that has been advanced by all the critics of Darwinism; but has been presented with most ability and force by the Duke of Argyll. I will therefore state it in his words.

If the doctrine of evolution be true--that is to say, if all organic creatures have been developed by ordinary generation from parents--then it follows of necessity that the prim 鑑 al germs must have contained potentially the whole succeeding series. Moreover, if that series has been developed gradually and very slowly, it follows, also as a matter of necessity, that every modification of structure must have been functionless at first, when it began to appear.... Things cannot be selected until they have first been produced. Nor can any structure be selected by utility in the struggle for existence until it has not only been produced, but has been so far perfected as to actually be used.

The Duke proceeds to argue that all adaptive structures must therefore originally have been due to special design: in the earlier stages of their development they must all have been what he calls "prophetic germs." Not yet themselves of any use, and therefore not yet capable of being improved by natural selection, both in their origin and in the first stages (at all events) of their development, they must be regarded as intentionally preparatory to the various uses which they subsequently acquire.

Now this argument, forcible as it appears at first sight, is really at fault both in its premiss and in its conclusion. By which I mean that, in the first place the premiss is not true, and, in the next place, that even if it were, the conclusion would not necessarily follow. The premiss is, "that every modification of structure must have been functionless at first, when it began to appear;" and the conclusion is, that, qu?functionless, such a modification cannot have been caused by natural selection. I will consider these two points separately.

First as to the premiss, it is not true that every modification of structure must necessarily be functionless when it first begins to appear. There are two very good reasons why such should not be the case in all instances, even if it should be the case in some. For, as a matter of observable fact, a very large proportional number of incipient organs are useful from the very moment of their inception. Take, for example, what is perhaps the most wonderful instance of refined mechanism in nature--the eye of a vertebrated animal. Comparative anatomy and embryology combine to testify that this organ had its origin in modifications of the endings of the ordinary nerves of the skin. Now it is evident that from the very first any modification of a cutaneous nerve whereby it was rendered able, in however small a degree, to be differently affected by light and by darkness would be of benefit to the creature presenting it; for the creature would thus be able to seek the one and shun the other according to the requirements of its life. And being thus useful from the very moment of its inception, it would afterwards be gradually improved as variations of more and more utility presented themselves, until not only would finer and finer degrees of difference between light and shade become perceptible, but even the outlines of solid bodies would begin to be appreciated. And so on, stage by stage, till from an ordinary nerve-ending in the skin is evolved the eye of an eagle.

Moreover, in this particular instance there is very good reason to suppose that the modification of the cutaneous nerves in question began by a progressive increase in their sensitiveness to temperature. Wherever dark pigment happened to be deposited in the skin--and we know that in all animals it is apt to be deposited in points and patches, as it were by accident, or without any "prophecy" as to future uses,--the cutaneous nerves in its vicinity would be better able to appreciate the difference between sun and shade in respect of temperature, even though as yet there were no change at all in these cutaneous nerves tending to make them responsive to light. Now it is easy to see how, from such a purely accidental beginning, natural selection would have had from the first sufficient material to act upon. It being of advantage to a lowly creature that it should distinguish with more and more delicacy, or with more and more rapidity, between light and darkness by means of its thermal sensations, the pigment spots in the skin would be rendered permanent by natural selection, while the nerves in that region would by the same agency be rendered more and more specialized as organs adapted to perceive changes of temperature, until from the stage of responding to the thermal rays of the non-luminous spectrum alone, they become capable of responding also to luminous.

So much, then, for the first consideration which serves to invalidate the Duke's premiss. The second consideration is, that very often an organ which began by being useful for the performance of one function, after having been fully developed for the performance of that function, finds itself, so to speak, accidentally fitted to the performance of some other and even more important function, which it thereupon begins to discharge, and so to undergo a new course of adaptive development. In such cases, and so far as the new function is concerned, the difficulty touching the first inception of an organ does not apply; for here the organ has already been built up by natural selection for one purpose, before it begins to discharge the other. As an example of such a case we may take the lung of an air-breathing animal. Originally the lung was a swim-bladder, or float, and as such it was of use to the aquatic ancestors of terrestrial animals. But as these ancestors gradually became more and more amphibious in their habits, the swim-bladder began more and more to discharge the function of a lung, and so to take a wholly new point of departure as regards its developmental history. But clearly there is here no difficulty with regard to the inception of its new function, because the organ was already well developed for one purpose before it began to serve another. Or, to take only one additional example, there are few structures in the animal kingdom so remarkable in respect of adaptation as is the wing of a bird or a bat; and at first sight it might well appear that a wing could be of no conceivable use until it had already acquired enormous proportional dimensions, as well as an immense amount of special elaboration as to its general form, size of muscle, amount of blood-supply, and so on. For, obviously, not until it had attained all these things could it even begin to raise the animal in the air. But observe how fallacious is this argument. Although it is perfectly true that a wing could be of no use as a wing until sufficiently developed to serve the purpose of flight, this is merely to say that until it has become a wing it is no use as a wing. It does not, however, follow that on this account it was of no prior use for any other purpose. The first modifications of the fore-limb which ended in its becoming an organ of flight may very well have been due to adapting it as an organ for increased rapidity of locomotion of other kinds--whether on land as in the case of its now degenerated form in the ostrich, or in water as in the case of the expanded fins of fish. Indeed, we may see the actual process of transition from the one function to the other in the case of "flying-fish." Here the progressive expansion of the pectoral fins must certainly have been always of use for continuously promoting rapidity of locomotion through water; and thus natural selection may have continuously increased their development until they now begin to serve also as wings for carrying the animal a short distance through air. Again, in the case of the so-called flying squirrels we find the limbs united to the body by means of large extensions of the skin, so-that when jumping from one tree to another the animal is able to sustain itself through a long distance in the air by merely spreading out its limbs, and thus allowing the skin-extensions to act after the manner of a parachute. Here, of course, we have not yet got a wing, any more than we have in the case of the flying-fish; but we have the foundations laid for the possible development of a future wing, upon a somewhat similar plan as that which has been so wonderfully perfected in the case of bats. And through all the stages of progressive expansion which the skin of the squirrel has undergone, the expansion has been of use, even though it has not yet so much as begun to acquire the distinctive functions of a wing. Here, then, there is obviously nothing "prophetic" in the matter, any more than there was in the case of the swim-bladder and the lung, or in that of the nerve-ending and the eye. In short, it is the business of natural selection to secure the highest available degree of adaptation for the time being; and, in doing this, it not unfrequently happens that an extreme development of a structure in one direction (produced by natural selection for the sake of better and better adapting the structure to perform some particular function) ends by beginning to adapt it to the performance of some other function. And, whenever this happens to be the case, natural selection forthwith begins to act upon the structure, so to speak, from a new point of departure.

So much, then, for the Duke's premiss--namely, that "every modification of structure must have been functionless at first, when it began to appear." This premiss is clearly opposed to observable fact. But now, the second position is that, even if this were not so, the Duke's conclusion would not follow. This conclusion, it will be remembered, is, that if incipient structures are useless, it necessarily follows that natural selection can have had no part whatever in their inception. Now, this is a conclusion which does not "necessarily" follow. Even if it be granted that there are structures which in their first beginnings are not of any use at all for any purpose, it is still possible that they may owe their origin to natural selection--not indeed directly, but indirectly. This possibility arises from the occurrence in nature of a principle which has been called the Correlation of Growth.

Mr. Darwin, who has paid more attention to this matter than any other writer, has shown, in considerable detail, that all the parts of any given organism are so intimately bound together, or so mutually dependent upon each other, that when one part is caused to change by means of natural selection, some other parts are very likely to undergo modification as a consequence. For example, there are several kinds of domesticated pigeons and fowls, which grow peculiar wing-like feathers on the feet. These are quite unlike all the other feathers in the animal, except those of the wing, to which they bear a very remarkable resemblance. Mr. Darwin records the case of a bantam where these wing-like feathers were nine inches in length, and I have myself seen a pigeon where they reproduced upon the feet a close imitation of the different kinds of feathers which occupy homologous positions in the wing--primaries, secondaries, and tertiaries all being distinctly repeated in their proper anatomical relations. Furthermore, in this case, as in most cases where such wing-feathers occur upon the feet, the third and fourth toes were partly united by skin; and, as is well known, in the wing of a bird the third and fourth digits are completely united by skin; "so that in feather-footed pigeons, not only does the exterior surface support a row of long feathers, like wing-feathers [which, as just stated, may in some cases be obviously differentiated into primaries, secondaries and tertiaries], but the very same digits which in the wing are completely united by skin become partially united by skin in the feet; and thus by the law of correlated variation of homologous parts, we can understand the curious connexion of feathered legs and membrane between the two outer toes[46]." The illustration is drawn from the specimen to which I have referred.

Many similar instances of the same law are to be met with throughout organic nature; and it is evident that in this principle we find a conceivable explanation of the origin of such adaptive structures as could not have been originated by natural selection acting directly upon themselves: they may have been originated by natural selection developing other adaptive structures elsewhere in the organism, the gradual evolution of which has entailed the production of these by correlation of growth. And, if so, when once started in this way, these structures, because thus accidentally useful, will now themselves come under the direct action of natural selection, and so have their further evolution determined with or without the correlated association which first led to their inception.

Of course it must be understood that in thus applying the principle of correlated growth, to explain the origin of adaptive structures where it is impossible to explain such origin by natural selection having from the first acted directly upon these structures themselves, Darwinists do not suppose that in all--or even in most--cases of correlated growth the correlated structures are of use. On the contrary, it is well known that structures due to correlated growth are, as a rule, useless. Being only the by-products of adaptive changes going on elsewhere, in any given case the chances are against these correlated effects being themselves of any utilitarian significance; and, therefore, as a matter of fact, correlated growths appear to be usually meaningless from the point of view of adaptation. Still, on the doctrine of chances, it is to be expected that sometimes a change of structure which has thus been indirectly produced by correlation of growth might happen to prove useful for some purpose or another; and in as many cases as such indirectly produced structures do prove useful, they will straightway begin to be improved by the direct action of natural selection. In all such cases, therefore, we should have an explanation of the origin of such a structure, which is the only point that we are now considering.

I think, then, that all this effectually disposes of the doctrine of "prophetic germs." But, before leaving the subject, I should like to make one further statement of greater generality than any which I have hitherto advanced. This statement is, that we must remember how large a stock of meaningless structures are always being produced in the course of specific transmutations, not only by correlation of growth, which we have just been considering, but also by the direct action of external conditions, together with the constant play of all the many and complex forces internal to organisms themselves. In other words, important as the principle of correlation undoubtedly is, we must remember that even this is very far from being the only principle which is concerned in the origination of structures that may or may not chance to be useful. Therefore, it is not only natural selection when operating indirectly through the correlation of growth that is competent to produce new structures without reference to utility. In all the complex action and reaction of internal and external forces, new variations are perpetually arising without any reference to utility, either present or future. Among all this multitude of promiscuous variations, the chances must be that some percentage will prove of some service, either from the first moment of their appearance, or else after they have undergone some amount of development. Such development prior to utility may be due, either to correlation of growth, to the structure having previously performed some other function, as already explained, or else to a continued operation of the causes which were concerned in the first appearance of originally useless characters. In a series of chapters which will be devoted to the whole question of utility in the next volume, I shall hope to give very good reasons for concluding that useless characters are not only of highly frequent occurrence, but are due to a variety of other causes besides correlation of growth. And, if so, the possibility of originally useless characters happening in some cases to become, by increased development, useful characters, is correspondingly increased. Among a hundred varietal or specific characters which are directly produced in as many different species by a change of climate, for example, some five or six may be potentially useful: that is to say, characters thus adventitiously produced in an incipient form may only require to be further developed by a continuance of the same causes as first originated them, in order that some percentage of the whole number shall become of some degree of use. Those professed followers of Darwin, therefore, who without any reason--or, as it appears to me, against all reason--deny the possibility of useless specific characters in any case or in any degree (unless correlated with useful characters), are playing into the hands of Darwin's critics by indirectly countenancing the difficulty which we are now considering. For, if correlation of growth is unreasonably supposed to be the only possible cause of the origin of incipient structures which are not useful from the first moment of their inception, clearly the field is greatly narrowed as regards the occurrence of incipient characters sufficient in amount--and, still more, in constancy of appearance and persistency of transmission--to admit of furnishing material for the working of natural selection. But in the measure that incipient characters--whether varietal or specific--are recognised as not always or "necessarily" useful from the moment of their inception, and yet capable of being developed to a certain extent by the causes which first led to their occurrence, in that measure is this line of criticism closed. For of all the variations which thus occur, it is only those which afterwards prove of any use that are laid hold upon and wrought up by natural selection into adaptive structures, or working organs. And, therefore, what we see in organic nature is the net outcome of the development of all the happy chances. So it comes that the appearance presented by organic nature as a whole is that of a continual fulfilment of structural prophecies, when, in point of fact, if we had a similar record of all the other variations it would be seen that possibly not one such prophecy in a thousand is ever destined to be fulfilled.

* * * * *

Here, then, I feel justified in finally taking leave of the difficulty from the uselessness of incipient organs, as this difficulty has been presented, in varying degrees of emphasis, by the Duke of Argyll, Mr. Mivart, Professors N 鋑 eli, Bronn, Broca, Eimer, and, indeed, by all other writers who have hitherto advanced it. For, as thus presented, I think I have shown that it admits of being adequately met. But now, I must confess, to me individually it does appear that behind this erroneous presentation of the difficulty there lies another question, which is deserving of much more serious attention. For although it admits of being easily shown--as I have just shown--that the difficulty as ordinarily presented fails on account of its extravagance, the question remains whether, if stated with more moderation, a real difficulty might not be found to remain.

My quarrel with the conclusion, like my quarrel with the premiss, is due to its universality. By saying in the premiss that all incipient organs are necessarily useless at the time of their inception, these writers admit of being controverted by fact; and by saying in the conclusion that, if all incipient organs are useless, it necessarily follows that in no case can natural selection have been the cause of building up an organ until it becomes useful, they admit of being controverted by logic. For, even if the premiss were true in fact--namely, that all incipient organs are useless at the time of their inception,--it would not necessarily follow that in no case could natural selection build up a useless structure into a useful one; because, although it is true that in no case can natural selection do this by acting on a useless structure directly, it may do so by acting on the useless structure indirectly, through its direct action on some other part of the organism with which the useless structure happens to be correlated. Moreover, as I believe, and will subsequently endeavour to prove, there is abundant evidence to show that incipient characters are often developed to a large extent by causes other than natural selection (or apart from any reference to utility), with the result that some of them thus happen to become of use, when, of course, the supposed difficulty is at an end.

But although it is thus easy to dispose of both the propositions in question, on account of their universality, stated more carefully they would require, as I have said, more careful consideration. Thus, if it had been said that some incipient organs are presumably useless at the time of their inception, and that in some of these cases it is difficult, or impossible, to conceive how the principle of correlation, or any other principle hitherto suggested, can apply-- then the question would have been raised from the sphere of logical discussion to that of biological fact. And the new question thus raised would have to be debated, no longer on the ground of general or abstract principles, but on that of special or concrete cases. Now until within the last year or two it has not been easy to find such a special or concrete case--that is to say, a case which can be pointed to as apparently excluding the possibility of natural selection having had anything to do with the genesis of an unquestionably adaptive structure. But eventually such a case has arisen, and the Duke of Argyll has not been slow in perceiving its importance. This case is the electric organ in the tail of the skate. No sooner had Professor Cossar Ewart published an abstract of his first paper on this subject, than the Duke seized upon it as a case for which, as he said, he had long been waiting-- namely, the case of an adaptive organ the genesis of which could not possibly be attributed to natural selection, and must therefore be attributed to supernatural design. Now, I do not deny that he is here in possession of an admirable case--a case, indeed, so admirable that it almost seems to have been specially designed for the discomfiture of Darwinians. Therefore, in order to do it full justice, I will show that it is even more formidable than the Duke of Argyll has represented.

Electric organs are known to occur in several widely different kinds of fish-- such as the Gymnotus and Torpedo. Wherever these organs do occur, they perform the function of electric batteries in storing and discharging electricity in the form of more or less powerful shocks. Here, then, we have a function which is of obvious use to the fish for purposes both of offence and defence. These organs are everywhere composed of a transformation of muscular, together with an enormous development of nervous tissue; but inasmuch as they occupy different positions, and are also in other respects dissimilar in the different zoological groups of fishes where they occur, no difficulty can be alleged as to these analogous organs being likewise homologous in different divisions of the aquatic vertebrata.

Now, in the particular case of the skate, the organ is situated in the tail, where it is of a spindle-like form, measuring, in a large fish, about two feet in length by about an inch in diameter at the middle of the spindle. Although its structure is throughout as complex and perfect as that of the electric organ in Gymnotus or Torpedo, its smaller size does not admit of its generating a sufficient amount of electricity to yield a discharge that can be felt by the hand. Nevertheless, that it does discharge under suitable stimulation has been proved by Professor Burdon Sanderson by means of a telephone; for he found that every time he stimulated the animal its electrical discharge was rendered audible by the telephone. Here, then, the difficulty arises. For of what conceivable use is such an organ to its possessor? We can scarcely suppose that any aquatic animal is more sensitive to electric shocks than is the human hand; and even if such were the case, a discharge of so feeble a kind taking place in water would be short-circuited in the immediate vicinity of the skate itself. So there can be no doubt that such weak discharges as the skate is able to deliver must be wholly imperceptible alike to prey and to enemies. Yet for the delivery of such discharges there is provided an organ of such high peculiarity and huge complexity, that, regarded as a piece of living mechanism, it deserves to rank as at once the most extremely specialized and the most highly elaborated structure in the whole animal kingdom. Thousands of separately formed elements are ranged in row after row, all electrically insulated one from another, and packed away into the smallest possible space, with the obvious end, or purpose, of conspiring together for the simultaneous delivery of an electric shock. Nevertheless, the shock when delivered is, as we have just seen, too slight to be of any conceivable use to the skate. Therefore it appears impossible to suggest how this astonishing structure--much more astonishing, in my opinion, than the human eye or the human hand--can ever have been begun, or afterwards developed, by means of natural selection. For if it be not even yet of any conceivable use to its possessor, clearly thus far survival of the fittest can have had nothing to do with its formation. On the other hand, seeing that electric organs when of larger size, as in the Gymnotus and Torpedo, are of obvious use to their possessors, the facts of the case, so far as the skate is concerned, assuredly do appear to sanction the doctrine of "prophetic germs." The organ in the skate seems to be on its way towards becoming such an organ as we meet with in these other animals; and, therefore, unless we can show that it is now, and in all previous stages of its evolution has throughout been, of use to the skate, the facts do present a serious difficulty to the theory of natural selection, while they readily lend themselves to the interpretation of a disposing or fore-ordaining mind, which knows how to construct an electric battery by thus transforming muscular tissue into electric tissue, and is now actually in process of constructing such an apparatus for the prospective benefit of future creatures.

Should it be suggested that possibly the electric organ of the skate may be in process of degeneration, and therefore that it is now the practically functionless remnant of an organ which in the ancestors of the skate was of larger size and functional use--against so obvious a suggestion there lie the whole results of Professor Ewart's investigations, which go to indicate that the organ is here not in a stage of degeneration, but of evolution. For instance, in Raia radiata, it does not begin to be formed out of the muscular tissue until some time after the animal has left the egg-capsule, and assumed all the normal proportions (though not yet the size) of the adult creature. The organ, therefore, is one of the very latest to appear in the ontogeny of R. radiata; and, moreover, it does not attain its full development (i. e. not merely growth, but transforming of muscular fibres into electrical elements) till the fish attains maturity. Read in the light of embryology, these facts prove, (1) that the electric organ of R. radiata must be one of the very latest products of the animal's phylogeny; and, (2) that as yet, at all events, it has not begun to degenerate. But, if not, it must either be at a stand-still, or it must be in course of further evolution; and, whichever of these alternatives we adopt, the difficulty of accounting for its present condition remains. In this connexion also it is worth while to remark that the electric organ, even after it has attained its full development, continues its growth with the growth of the fish, and this in a much higher ratio, either than the tail alone, or the whole animal. Lastly, Prof. Burdon Sanderson finds that section for section the organ in the skate is as efficient as it is in Torpedo. It is evident that these facts also point to the skate's organ being in course of phylogenetic evolution.

Again, it cannot be answered that the principle of correlation may be drawn upon in mitigation of the difficulty. The structure of the electric organ is far too elaborate, far too specialized, and far too obviously directed to a particular end, to admit of our conceivably supposing it due to any accidental correlation with structural changes going on elsewhere. Even as regards the initial changes of muscle-elements into electrical-elements, I do not think the principle of correlation can be reasonably adduced by way of explanation; for, as shown in the illustrations, even this initial change is most extraordinarily peculiar, elaborate, and specialized. But, be this as it may, I am perfectly certain that the principle of correlation cannot possibly be adduced to explain the subsequent association of these electrical elements into an electric battery, actuated by a special nervous mechanism of enormous size and elaboration--unless of course, the progress of such a structure were assumed to have been throughout of some utility. Under this supposition, however, the principle of correlation would be forsaken in favour of that of natural selection; and we should again be in the presence of the same difficulty as that with which we started.

But now, and further, if we do thus abandon correlation in favour of natural selection, and therefore if for the sake of saving an hypothesis we assume that the organ as it now stands must be of some use to the existing skate, we should still have to face the question--Of what conceivable use can those initial stages of its formation have been, when first the muscle-elements began to be changed into the very different electrical-elements, and when therefore they became useless as muscles while not yet capable of performing even so much of the electrical function as they now perform?

Lastly, we must remember that not only have we here the most highly specialized, the most complex, and altogether the most elaboratively adaptive organ in the animal kingdom; but also that in the formation of this structure there has been needed an altogether unparalleled expenditure of the most physiologically expensive of all materials--namely, nervous tissue. Whether estimated by volume or by weight, the quantity of nervous tissue which is consumed in the electric organ of the skate is in excess of all the rest of the nervous system put together. It is needless to say that nowhere else in the animal kingdom--except, of course, in other electric fishes--is there any approach to so enormous a development of nervous tissue for the discharge of a special function. Therefore, as nervous tissue is, physiologically speaking, the most valuable of all materials, we are forced to conclude that natural selection ought strongly to have opposed the evolution of such organs, unless from the first moment of their inception, and throughout the whole course of their development, they were of some such paramount importance as biologically to justify so unexampled an expenditure. Yet this paramount importance does not admit of being so much as surmised, even where the organ has already attained the size and degree of elaboration which it presents in the skate.

In view of all these considerations taken together, I freely confess that the difficulty presented by this case appears to me of a magnitude and importance altogether unequalled by that of any other single case--or any series of cases--which has hitherto been encountered by the theory of natural selection. So that, if there were many other cases of the like kind to be met with in nature, I should myself at once allow that the theory of natural selection would have to be discarded. But inasmuch as this particular case stands so far entirely by itself, and therefore out of analogy with thousands, or even millions, of other cases throughout the whole range of organic nature, I am constrained to feel it more probable that the electric organ of the skate will some day admit of being marshalled under the general law of natural selection--in just the same way as proved to be the case with the conspicuous colouring of those caterpillars, which, as explained in the last chapter, at one time seemed to constitute a serious difficulty to the theory, and yet, through a better knowledge of all the relations involved, has now come to constitute one of the strongest witnesses in its favour.

* * * * *

I have now stated all the objections of any importance which have hitherto been brought against the theory of natural selection, excepting three, which I left to be dealt with together because they form a logically connected group. With a brief consideration of these, therefore, I will bring this chapter to a close.

The three objections to which I allude are, (1) that a large proportional number of specific, as well as of higher taxonomic characters, are seemingly useless characters, and therefore do not lend themselves to explanation by the Darwinian theory; (2) that the most general of all specific characters--viz. cross-infertility between allied species--cannot possibly be due to natural selection, as is demonstrated by Darwin himself; (3) that the swamping effects of free intercrossing must always render impossible by natural selection alone any evolution of species in divergent (as distinguished from serial) lines of change.

These three objections have been urged from time to time by not a few of the most eminent botanists and zoologists of our century; and from one point of view I cannot myself have the smallest doubt that the objections thus advanced are not only valid in themselves, but also by far the most formidable objections which the theory of natural selection has encountered. From another point of view, however, I am equally convinced that they all admit of absolute annihilation. This strong antithesis arises, as I have said, from differences of standpoint, or from differences in the view which we take of the theory of natural selection itself. If we understand this theory to set forth natural selection as the sole cause of organic evolution, then all the above objections to the theory are not merely, as already stated, valid and formidable, but as I will now add, logically insurmountable. On the other hand, if we take theory to consist merely in setting forth natural selection as a factor of organic evolution, even although we believe it to have been the chief factor or principal cause, all the three objections in question necessarily vanish. For in this case, even if it be satisfactorily proved that the theory of natural selection is unable to explain the three classes of facts above mentioned, the theory is not thereby affected: facts of each and all of these classes may be consistently left by the theory to be explained by causes other than natural selection--whether these be so far capable or incapable of hypothetical formulation. Thus it is evident that whether the three objections above named are to be regarded as logically insurmountable by the theory, or as logically non-existent in respect to it, depends simply upon the manner in which the theory itself is stated.

In the next volume a great deal more will have to be said upon these matters--especially with regard to the causes other than natural selection which in my opinion are capable of explaining these so-called "difficulties." In the present connexion, however, all I have attempted to show is, that, whatever may be thought touching the supplementary theories whereby I shall endeavour to explain the facts of inutility, cross-sterility, and non-occurrence of free intercrossing, no one of these facts is entitled to rank as an objection against the theory of natural selection, unless we understand this theory to claim an exclusive prerogative in the field of organic evolution. This, as we have previously seen, is what Mr. Wallace does claim for it; while on the other hand, Mr. Darwin expressly--and even vehemently--repudiates the claim: from which it follows that all the three main objections against the theory of natural selection are objections which vitally affect the theory only as it has been stated and upheld by Wallace. As the theory has been stated and upheld by Darwin, all these objections are irrelevant. This is a fact which I had not myself perceived at the time when I mentioned these objections in a paper entitled Physiological Selection, which was published in 1886. The discussions to which that paper gave rise, however, led me to consider these matters more closely; and further study of Darwin's writings, with these matters specially in view, has led me to see that none of the objections in question are relevant to his theory, as distinguished from that of Mr. Wallace. This, I acknowledge, I ought to have perceived before I published the paper just alluded to; but in those days I had had no occasion to follow out the differences between Darwin and Wallace to all their consequences, and therefore adopted the prevalent view that their theories of evolution were virtually identical. Now, however, I have endeavoured to make it clear that the points wherein they differ involve the important consequences above set forth. All these the most formidable objections against the theory of natural selection arise simply and solely from what I conceive to be the erroneous manner in which the theory has been presented by Darwin's distinguished colleague.

* * * * *

I have now considered, as impartially as I can, all the main criticisms and objections which have been brought against the theory of natural selection;

and the result is to show that, neither singly nor collectively, are they entitled to much weight. On the other hand, as we have seen in the preceding chapter, there is a vast accumulation of evidence in favour of the theory. Hence, it is no wonder that the theory has now been accepted by all naturalists, with scarcely any one notable exception, as at any rate the best working hypothesis which has ever been propounded whereby to explain the facts of organic evolution. Moreover, in the opinion of those most competent to judge, the theory is entitled to be regarded as something very much more than a working hypothesis: it is held to be virtually a completed induction, or, in other words, the proved exhibition of a general law, whereby the causation of organic evolution admits of being in large part--if not altogether--explained.

Now, whether or not we subscribe to this latter conclusion ought, I think, to depend upon what we mean by an explanation in the case which is before us. If we mean only that, given the large class of known facts and unknown causes which are conveniently summarized under the terms Heredity and Variability, then the further facts of Struggle and Survival serve, in some considerable degree or another, to account for the phenomena of adaptive evolution, I cannot see any room to question that the evidence is sufficient to prove the statement. But it is clear that by taking for granted these great facts of Heredity and Variability, we have assumed the larger part of the problem as a whole. Or, more correctly, by thus generalizing, in a merely verbal form, all the unknown causes which are concerned in these two great factors of the process in question, we are not so much as attempting to explain the precedent causation which serves as a condition to the process. Much more than half the battle would already have been won, had Darwin's predecessors been able to explain the causes of Heredity and Variation; hence it is but a very partial victory which we have hitherto gained in our recent discovery of the effects of Struggle and Survival.

Yet partial though it be in relation to the whole battle, in itself, or considered absolutely, there can be no reasonable doubt that it constitutes the greatest single victory which has ever been gained by the science of Biology. For this very reason, however, it behoves us to consider all the more carefully the extent to which it goes. But my discussion of this matter must be relegated to the next volume, where I hope to give abundant proof of the soundness of Darwin's judgment as conveyed in the words:--"I am convinced that natural selection has been the main, but not the exclusive, means of modification."

CHAPTER X.

THE THEORY OF SEXUAL SELECTION, AND CONCLUDING REMARKS.

Although the explanatory value of the Darwinian theory of natural selection is, as we have now seen, incalculably great, it nevertheless does not meet those phenomena of organic nature which perhaps more than any other attract the general attention, as well as the general admiration, of mankind: I mean all that class of phenomena which go to constitute the Beautiful. Whatever value beauty as such may have, it clearly has not a life-preserving value. The gorgeous plumage of a peacock, for instance, is of no advantage to the peacock in his struggle for life, and therefore cannot be attributed to the agency of natural selection. Now this fact of beauty in organic structures is a fact of wide generality--almost as wide, indeed, as is the fact of their utility. Mr. Darwin, therefore, suggested another hypothesis whereby to render a scientific explanation of this fact. Just as by his theory of natural selection he sought to explain the major fact of utility, so did he endeavour to explain the minor fact of beauty by a theory of what he termed Sexual Selection.

It is a matter of observation that the higher animals do not pair indiscriminately; but that the members of either sex prefer those individuals of the opposite sex which are to them most attractive. It is important to understand in limine that nobody has ever attempted to challenge this statement. In other words, it is an unquestionable fact that among many of the higher animals there literally and habitually occurs a sexual selection; and this fact is not a matter of inference, but, as I have said, a matter of observation. The inference only begins where, from this observable fact, it is argued,--1st, that the sexual selection has reference to an aesthetic taste on the part of the animals themselves; and 2nd, that, supposing the selection to be determined by such a taste, the cause thus given is adequate to explain the phenomena of beauty which are presented by these animals. I will consider these two points separately.

From the evidence which Darwin has collected, it appears to me impossible to doubt that an aesthetic sense is displayed by many birds, and not a few mammals. This of course does not necessarily imply that the standards of such a sense are the same as our own; nor does it necessarily imply that there is any constant relation between such a sense and high levels of intelligence in other respects. In point of fact, such is certainly not the case, because the best evidence that we have of an aesthetic sense in animals is derived from birds, and not from mammals. The most cogent cases to quote in this connexion are those of the numerous species of birds which habitually adorn their nests with gaily coloured feathers, wool, cotton, or any other gaudy materials which they may find lying about the woods and fields. In many cases a marked preference is shown for particular objects--as, for instance, in the case of the Syrian nut-hatch, which chooses the iridescent wings of insects, or that of the great crested fly-catcher, which similarly chooses the cast-off skins of snakes. But no doubt the most remarkable of these cases is that of the baya-bird of Asia, which after having completed its bottle-shaped and chambered nest[47], studs it over with small lumps of clay, both inside and out, upon which the cock-bird sticks fire-flies, apparently for the sole purpose of securing a brilliantly decorative effect. Other birds, such as the hammer-head of Africa, adorn the surroundings of their nests (which are built upon the ground) with shells, bones, pieces of broken glass and earthenware, or any objects of a bright and conspicuous character which they may happen to find. The most consummate artists in this respect are, however, the bower-birds; for the species of this family construct elaborate play-houses in the form of arched tunnels, built of twigs upon the ground. Through and around such a tunnel they chase one another; and it is always observable that not only is the floor paved with a great collection of shells, bones, coloured stones, and any other brilliant objects which they are able to carry in their beaks, but also that the walls are decorated with the most gaudy articles which the birds can find. There is one genus, in Papua, which even goes so far as to provide the theatre with a surrounding garden. A level piece of ground is selected as a site for the building. The latter is about two feet high, and constructed round the growing stalk of a shrub, which therefore serves as a central pillar to which the frame-work of the roof is attached. Twigs are woven into this frame-work until the whole is rendered rain-proof. The tent thus erected is about nine feet in circumference at its base, and presents a large arch as an entrance. The central pillar is banked up with moss at its base, and a gallery is built round the interior of the edifice. This gallery is decorated with flowers, fruits, fungi, &c. These are also spread over the garden, which covers about the same area as the play-house. The flowers are said to be removed when they fade, while fresh ones are gathered to supply their places. Thus the garden is always kept bright with flowers, as well as with the brilliant green of mosses, which are collected and distributed in patches, resembling tiny lawns.

[47] The chambers are three in number. The two upper ones are occupied respectively by the male and the sitting female. The lower one serves as a general living room when the young are hatched.

Now these sundry cases alone seem to prove a high degree of the aesthetic sense as occurring among birds; for, it is needless to say, none of the facts just mentioned can be due to natural selection, seeing that they have no reference to utility, or the preservation of life. But if an aesthetic sense occurs in birds, we should expect, on a priori grounds, that it would probably be exercised with reference to the personal appearance of the sexes. And this expectation is fully realized. For it is an observable fact that in most species of birds where the males are remarkable for the brilliancy of their plumage, not only is this brilliancy most remarkable during the pairing season, but at this season also the male birds take elaborate pains to display their charms before the females. Then it is that the peacock erects his tail to strut round and round the hens, taking care always to present to them a front view, where the coloration is most gorgeous. And the same is true of all other gaily coloured male birds. During the pairing season they actively compete with one another in exhibiting their attractiveness to the females; and in many cases there are added all sorts of extraordinary antics in the way of dancings and crowings. Again, in the case of all song-birds, the object of the singing is to please the females; and for this purpose the males rival one another to the best of their musical ability.

Thus there can be no question that the courtship of birds is a highly elaborate business, in which the males do their best to surpass one another in charming the females. Obviously the inference is that the males do not take all this trouble for nothing; but that the females give their consent to pair with the males whose personal appearance, or whose voice, proves to be the most attractive. But, if so, the young of the male bird who is thus selected will inherit his superior beauty; and thus, in successive generations, a continuous advance will be made in the beauty of plumage or of song, as the case may be,--both the origin and development of beauty in the animal world being thus supposed due to the aesthetic taste of animals themselves.

Such is the theory of sexual selection in its main outlines; and with regard to it we must begin by noting two things which are of most importance. In the first place, it is a theory wholly and completely distinct from the theory of natural selection; so that any truth or error in the one does not in the least affect the other. The second point is, that there is not so great a wealth of evidence in favour of sexual selection as there is in favour of natural selection; and, therefore, that while all naturalists nowadays accept natural selection as a (whether or not the) cause of adaptive, useful, or life-preserving structures, there is no such universal--but only a very general--agreement with reference to sexual selection as a cause of decorative, beautiful, or life-embellishing structures. Nevertheless, the evidence in favour of sexual selection is both large in amount and massive in weight.

Our consideration of this evidence will bring us to the second division of our subject, as previously marked out for discussion--namely, granting that an aesthetic sense occurs in certain large divisions of the animal kingdom, what is the proof that such a sense is a cause of the beauty which is presented by the animals in question?

Before proceeding to state this proof, however, it is desirable to observe that under the theory of sexual selection Darwin has included two essentially different classes of facts. For besides the large class of facts to which I have thus far been alluding,--i. e. the cases where two sexes of the same species differ from one another in respect of ornamentation,--there is another class of facts equally important, namely, the cases where the two sexes of the same species differ from one another in respect of size, strength, and the possession of natural weapons, such as spurs, horns, &c. In most of these cases it is the males which are thus superiorly endowed; and it is a matter of observation that in all cases where they are so endowed they use their superior strength and natural weapons for fighting together, in order to secure possession of the females. Hence results what Mr. Darwin has called the Law of Battle between males of the same species; and this law of battle he includes under his theory of sexual selection. But it is evident that the principle which is operative in the law of battle differs from the principle which is concerned in the form of sexual selection that has to do with embellishment, and consequent charm. The law of battle, in fact, more nearly approaches the law of natural selection; seeing that it expresses the natural advantages of brute force in the struggling of rival animals, and so frequently results in death of the less fitted, as distinguished from a mere failure to propagate. Now against this doctrine of the law of battle, and the consequences to which it leads in the superior fighting powers of male animals, no objection has been raised in any quarter. It is only with regard to the other aspect of the theory of sexual selection--or that which is concerned with the superior embellishment of male animals--that any difference of opinion obtains. I will now proceed to give the main arguments on both sides of this question, beginning with a summaryof the evidences in favour of sexual selection.

In the first place, the fact that secondary sexual characters of the embellishing kind are so generally restricted to the male sex in itself seems to constitute very cogent proof that, in some way or another, such characters are connected with the part which is played by the male in the act of propagation. Moreover, secondary sexual characters of this kind are of quite as general occurrence as are those of the other kind which have to do with rivalry in battle; and the former are usually of the more elaborate description. Therefore, as there is no doubt that secondary sexual characters of the one order have an immediate purpose to serve in the act of propagation, we are by this close analogy confirmed in our surmise that secondary sexual characters of the other, and still more elaborate, order are likewise so concerned. Moreover, this view of their meaning becomes still further strengthened when we take into consideration the following facts. Namely, (a) secondary sexual characters of the embellishing kind are, as a rule, developed only at maturity; and most frequently during only a part of the year, which is invariably the breeding season: (b) they are always more or less seriously affected by emasculation: (c) they are always, and only, displayed in perfection during the act of courtship: (d) then, however, they are displayed with the most elaborate pains; yet always, and only, before the females: (e) they appear, at all events in many cases, to have the effect of charming the females into a performance of the sexual act; while it is certain that in many cases, both among quadrupeds and birds, individuals of the one sex are capable of feeling a strong antipathy against, or a strong preference for, certain individuals of the opposite sex.

Such are the main lines of evidence in favour of the theory of sexual selection. And although it is enough that some of them should be merely stated as above in order that their immense significance should become apparent, in the case of others a bare statement is not sufficient for this purpose. More especially is this the case as regards the enormous profusion, variety, and elaboration of sexually-embellishing characters which occur in birds and mammals--not to mention several divisions of Arthropoda; together with the extraordinary amount of trouble which, in a no less extraordinary number of different ways, is taken by the male animals to display their embellishments before the females. And even in many cases where to our eyes there is no particular embellishment to display, the process of courtship consists in such an elaborate performance of dancings, struttings, and attitudinizings that it is scarcely possible to doubt their object is to incite the opposite sex. Here, for instance, is a series of drawings illustrating the courtship of spiders. I choose this case as an example, partly because it is the one which has been published most recently, and partly because it is of particular interest as occurring so low down in the zoological scale. I am indebted to the kindness of Mr. and Mrs. Peckham for permission to reproduce these few selected drawings from their very admirable work, which is published by the Natural History Society of Wisconsin, U.S. It is evident at a glance that all these elaborate, and to our eyes ludicrous, performances are more suggestive of incitation than of any other imaginable purpose. And this view of the matter is strongly corroborated by the fact that it is the most brightly coloured parts of the male spiders which are most obtruded upon the notice of the female by these peculiar attitudes--in just the same way as is invariably the case in the analogous phenomena of courtship among birds, insects, &c.

But so great is the mass of material which Darwin has collected in proof of all the points mentioned in the foregoing paragraph, that to attempt anything in the way of an epitome would really be to damage its evidential force. Therefore I deem it best simply to refer to it as it stands in his Descent of Man, concluding, as he concludes,--"This surprising uniformity in the laws regulating the differences between the sexes in so many and such widely separated classes is intelligible if we admit the action throughout all the higher divisions of the animal kingdom of one common cause, namely, sexual selection"; while, as he might well have added, it is difficult to imagine that all the large classes of facts which an admission of this common cause serves to explain, can ever admit of being rendered intelligible by any other theory.

We may next proceed to consider the objections which have been brought against the theory of sexual selection. And this is virtually the same thing as saying that we may now consider Mr. Wallace's views upon the subject.

Reserving for subsequent consideration the most general of these objections--namely, that at best the theory can only apply to the more intelligent animals, and so must necessarily fail to explain the phenomena of beauty in the less intelligent, or in the non-intelligent, as well as in all species of plants--we may take seriatim the other objections which, in the opinion of Mr. Wallace, are sufficient to dispose of the theory even as regards the higher animals.

In the first place, he argues that the principal cause of the greater brilliancy of male animals in general, and of male birds in particular, is that they do not so much stand in need of protection arising from concealment as is the case with their respective females. Consequently natural selection is not so active in repressing brilliancy of colour in the males, or, which amounts to the same thing, is more active in "repressing in the female those bright colours which are normally produced in both sexes by general laws."

Next, he argues that not only does natural selection thus exercise a negative influence in passively permitting more heightened colour to appear in the males, but even exercises a positive influence in actively promoting its development in the males, while, at the same time, actively repressing its appearance in the females. For heightened colour, he says, is correlated with health and vigour; and as there can be no doubt that healthy and vigorous birds best provide for their young, natural selection, by always placing its premium on health and vigour in the males, thus also incidentally promotes, through correlated growth, their superior coloration.

Again, with regard to the display which is practised by male birds, and which constitutes the strongest of all Mr. Darwin's arguments in favour of sexual selection, Mr. Wallace points out that there is no evidence of the females being in any way affected thereby. On the other hand, he argues that this display may be due merely to general excitement; and he lays stress upon the more special fact that moveable feathers are habitually erected under the influence of anger and rivalry, in order to make the bird look more formidable in the eyes of antagonists.

Furthermore, he adduces the consideration that, even if the females are in any way affected by colour and its display on the part of the males, and if, therefore, sexual selection be conceded a true principle in theory, still we must remember that, as a matter of fact, it can only operate in so far as it is allowed to operate by natural selection. Now, according to Mr. Wallace, natural selection must wholly neutralize any such supposed influence of sexual selection. For, unless the survivors in the general struggle for existence happen to be those which are also the most highly ornamented, natural selection must neutralize and destroy any influence that may be exerted by female selection. But obviously the chances against the otherwise best fitted males happening to be likewise the most highly ornamented must be many to one, unless, as Wallace supposes, there is some correlation between embellishment and general perfection, in which case, as he points out, the theory of sexual selection lapses altogether, and becomes but a special case of natural selection.

Once more, Mr. Wallace argues that the evidence collected by Mr. Darwin himself proves that each bird finds a mate under any circumstances--a general fact which in itself must quite neutralize any effect of sexual selection of colour or ornament, since the less highly coloured birds would be at no disadvantage as regards the leaving of healthy progeny.

Lastly, he urges the high improbability that through thousands of generations all the females of any particular species--possibly spread over an enormous area--should uniformly and always have displayed exactly the same taste with respect to every detail of colour to be presented by the males.

Now, without any question, we have here a most powerful array of objections against the theory of sexual selection. Each of them is ably developed by Mr. Wallace himself in his work on Tropical Nature; and although I have here space only to state them in the most abbreviated of possible forms, I think it will be apparent how formidable these objections appear. Unfortunately the work in which they are mainly presented was published several years after the second edition of the Descent of Man, so that Mr. Darwin never had a suitable opportunity of replying. But, if he had had such an opportunity, as far as I can judge it seems that his reply would have been more or less as follows.

In the first place, Mr. Wallace fails to distinguish between brilliancy and ornamentation--or between colour as merely "heightened," and as distinctively decorative. Yet there is obviously the greatest possible difference between these two things. We may readily enough admit that a mere heightening of already existing coloration is likely enough--at all events in many cases--to accompany a general increase of vigour, and therefore that natural selection, by promoting the latter, may also incidentally promote the former, in cases where brilliancy is not a source of danger. But clearly this is a widely different thing from showing that not only a general brilliancy of colour, but also the particular disposition of colours, in the form of ornamental patterns, can thus be accounted for by natural selection. Indeed, it is expressly in order to account for the occurrence of such ornamental patterns that Mr. Darwin constructed his theory of sexual selection; and therefore, by thus virtually ignoring the only facts which that theory endeavours to explain, Mr. Wallace is not really criticizing the theory at all. By representing that the theory has to do only with brilliancy of colour, as distinguished from disposition of colours, he is going off upon a false issue which has never really been raised[48]. Look, for example, at a peacock's tail. No doubt it is sufficiently brilliant; but far more remarkable than its brilliancy is its elaborate pattern on the one hand, and its enormous size on the other. There is no conceivable reason why mere brilliancy of colour, as an accidental concomitant of general vigour, should have run into so extraordinary, so elaborate, and so beautiful a design of colours. Moreover, this design is only unfolded when the tail is erected, and the tail is not erected in battle (as Mr. Wallace's theory of the erectile function in feathers would require), but in courtship; obviously, therefore, the purpose of the pattern, so to speak, is correlated with the act of courtship--it being only then, in fact, that the general purpose of the whole structure, as well as the more special purpose of the pattern, becomes revealed. Lastly, the fact of this whole structure being so large, entailing not only a great amount of physiological material in its production, but also of physiological energy in carrying about such a weight, as well as of increased danger from impeding locomotion and inviting capture--all this is obviously incompatible with the supposition of the peacock's tail having been produced by natural selection. And such a case does not stand alone. There are multitudes of other instances of ornamental structures imposing a drain upon the vital energies of their possessors, without conferring any compensating benefit from a utilitarian point of view. Now, in all these cases, without any exception, such structures are ornamental structures which present a plain and obvious reference to the relationship of the sexes. Therefore it becomes almost impossible to doubt-- first, that they exist for the sake of ornament; and next, that the ornament exists on account of that relationship. If such structures were due merely to a superabundance of energy, as Mr. Wallace supposes, not only ought they to have been kept down by the economizing influence of natural selection; but we can see no reason, either why they should be so highly ornamental on the one hand, or so exclusively related to the sexual relationship on the other.

[48] Note C.

Finally, we must take notice of the fact that where peculiar structures are concerned for purposes of display in courtship, the elaboration of these structures is often no less remarkable than that of patterns where colours are thus concerned. Take, for example, the case of the Bell-bird, which I select from an innumerable number of instances that might be mentioned because, while giving a verbal description of this animal, Darwin does not supply a pictorial representation thereof. The bird, which lives in South America, has a very loud and peculiar call, that can be heard at a distance of two or three miles. The female is dusky-green; but the adult male is a beautiful white, excepting the extraordinary structure with which we are at present concerned. This is a tube about three inches long, which rises from the base of the beak. It is jet black, and dotted over with small downy feathers. The tube is closed at the top, but its cavity communicates with the palate, and thus the whole admits of being inflated from within, when, of course, it stands erect as represented in one of the two drawings. When not thus inflated, it hangs down, as shown in the second figure, which represents the plumage of a young male. (Fig. 124.)

In another species of the genus there are three of these appendages--the two additional ones being mounted on the corners of the mouth. (Fig. 125.) In all species of the genus (four in number) the tubes are inflated during courtship, and therefore perform the function of sexual embellishments. Now the point to which I wish to draw attention is, that so specialized and morphologically elaborate a structure cannot be regarded as merely adventitious. It must have been developed by some definite cause, acting through a long series of generations. And as no other function can be assigned to it than that of charming the female when it is erected in courtship, the peculiarity of form and mechanism which it presents--like the elaboration of patterns in cases where colour only is concerned--virtually compels us to recognise in sexual selection the only conceivable cause of its production.

For these reasons I think that Mr. Wallace's main objection falls to the ground. Passing on to his subsidiary objections, I do not see much weight in his merely negative difficulty as to there being an absence of evidence upon hen birds being charmed by the plumage, or the voice, of their consorts. For, on the one hand, it is not very safe to infer what sentiments may be in the mind of a hen; and, on the other hand, it is impossible to conceive what motive can be in the mind of a cock, other than that of making himself attractive, when he performs his various antics, displays his ornamental plumes, or sings his melodious songs. Considerations somewhat analogous apply to the difficulty of supposing so much similarity and constancy of taste on the part of female animals as Mr. Darwin's theory undoubtedly requires. Although we know very little about the psychology of the lower animals, we do observe in many cases that small details of mental organization are often wonderfully constant and uniform throughout all members of a species, even where it is impossible to suggest any utility as a cause.

Again, as regards the objection that each bird finds a mate under any circumstances, we have here an obvious begging of the whole question. That every feathered Jack should find a feathered Jill is perhaps what we might have antecedently expected; but when we meet with innumerable instances of ornamental plumes, melodious songs, and the rest, as so many witnesses to a process of sexual selection having always been in operation, it becomes irrational to exclude such evidence on account of our antecedent prepossessions.

There remains the objection that the principles of natural selection must necessarily swallow up those of sexual selection. And this consideration, I doubt not, lies at the root of all Mr. Wallace's opposition to the supplementary theory of sexual selection. He is self-consistent in refusing to entertain the evidence of sexual selection, on the ground of his antecedent persuasion that in the great drama of evolution there is no possible standing-ground for any other actor than that which appears in the person of natural selection. But here, again, we must refuse to allow any merely antecedent presumption to blind our eyes to the actual evidence of other agencies having co-operated with natural selection in producing the observed results. And, as regards the particular case now before us, I think I have shown, as far as space will permit, that in the phenomena of decorative colouring (as distinguished from merely brilliant colouring), of melodious song (as distinguished from merely tuneless cries), of enormous arborescent antlers (as distinguished from merely offensive weapons), and so forth--I say that in all these phenomena we have phenomena which cannot possibly be explained by the theory of natural selection; and, further, that if they are to be explained at all, this can only be done, so far as we can at present see, by Mr. Darwin's supplementary theory of sexual selection.

I have now briefly answered all Mr. Wallace's objections to this supplementary theory, and, as previously remarked, I feel pretty confident that, at all events in the main, the answer is such as Mr. Darwin would himself have supplied, had there been a third edition of his work upon the subject. At all events, be this as it may, we are happily in possession of unquestionable evidence that he believed all Mr. Wallace's objections to admit of fully satisfactory answers. For his very last words to science--read only a few hours before his death at a meeting of the Zoological Society-- were:

I may perhaps be here permitted to say that, after having carefully weighed, to the best of my ability, the various arguments which have been advanced against the principle of sexual selection, I remain firmly convinced of its truth[49].

[49] Since the above exposition of the theory of sexual selection was written, Mr. Poulton has published his work on the Colours of Animals. He there reproduces some of the illustrations which occur in Mr. and Mrs. Peckham's work on Sexual Selection in Spiders, and furnishes appropriate descriptions. Therefore, while retaining the illustrations, I have withdrawn my own descriptions.

Mr. Poulton has also in his book supplied a summary of the arguments for and against the theory of sexual selection in general. Of course in nearly all respects this corresponds with the summary which is given in the foregoing pages; but I have left the latter as it was originally written, because all the critical part is reproduced verbatim from a review of Mr. Wallace's Darwinism, of a date still earlier than that of Mr. Poulton's book--viz. Contemporary Review, August, 1889.

Concluding Remarks.

I will now conclude this chapter, and with it the present volume, by offering a few general remarks on what may be termed the philosophical relations of Darwinian doctrine to the facts of adaptation on the one hand, and to those of beauty on the other. Of course we are all aware that before the days of this doctrine the facts of adaptation in organic nature were taken to constitute the clearest possible evidence of special design, on account of the wonderful mechanisms which they everywhere displayed; while the facts of beauty were taken as constituting no less conclusive evidence of the quality of such special design as beneficent, not to say artistic. But now that the Darwinian doctrine appears to have explained scientifically the former class of facts by its theory of natural selection, and the latter class of facts by its theory of sexual selection, we may fitly conclude this brief exposition of the doctrine as a whole by considering what influence such naturalistic explanations may fairly be taken to exercise upon the older, or super-naturalistic, interpretations.

To begin with the facts of adaptation, we must first of all observe that the Darwinian doctrine is immediately concerned with these facts only in so far as they occur in organic nature. With the adaptations--if they can properly be so called--which occur in all the rest of nature, and which go to constitute the Cosmos as a whole so wondrous a spectacle of universal law and perfect order, this doctrine is but indirectly concerned. Nevertheless, it is of course fundamentally concerned with them to the extent that it seeks to bring the phenomena of organic nature into line with those of inorganic; and therefore to show that whatever view we may severally take as to the kind of causation which is energizing in the latter we must now extend to the former. This is usually expressed by saying that the theory of evolution by natural selection is a mechanical theory. It endeavours to comprise all the facts of adaptation in organic nature under the same category of explanation as those which occur in inorganic nature--that is to say, under the category of physical, or ascertainable, causation. Indeed, unless the theory has succeeded in doing this, it has not succeeded in doing anything--beyond making a great noise in the world. If Mr. Darwin has not discovered a new mechanical cause in the selection principle, his labour has been worse than in vain.

Now, without unduly repeating what has already been said in Chapter VIII, I may remark that, whatever we may each think of the measure of success which has thus far attended the theory of natural selection in explaining the facts of adaptation, we ought all to agree that, considered as a matter of general reasoning, the theory does certainly refer to a vera causa of a strictly physical kind; and, therefore, that no exception can be taken to the theory in this respect on grounds of logic. If the theory in this respect is to be attacked at all, it can only be on grounds of fact--namely, by arguing that the cause does not occur in nature, or that, if it does, its importance has been exaggerated by the theory. Even, however, if the latter proposition should ever be proved, we may now be virtually certain that the only result would be the relegation of all the residual phenomena of adaptation to other causes of the physical order--whether known or unknown. Hence, as far as the matter of principle is concerned, we may definitely conclude that the great naturalistic movement of our century has already brought all the phenomena of adaptation in organic nature under precisely the same category of mechanical causation, as similar movements in previous centuries have brought all the known phenomena of inorganic nature: the only question that remains for solution is the strictly scientific question touching the particular causes of the mechanical order which have been at work.

So much, then, for the phenomena of adaptation. Turning next to those of beauty, we have already seen that the theory of sexual selection stands to these in precisely the same relation as the theory of natural selection does to those of adaptation. In other words, it supplies a physical explanation of them; because, as far as our present purposes are concerned, it may be taken for granted, or for the sake of argument, that inasmuch as psychological elements enter into the question the cerebral basis which they demand involves a physical side.

There is, moreover, this further point of resemblance between the two theories: neither of them has any reference to inorganic nature. Therefore, with the charm or the loveliness of landscapes, of earth and sea and sky, of pebbles, crystals, and so forth, we have at present nothing to do. How it is that so many inanimate objects are invested with beauty--why it is that beauty attaches to architecture, music, poetry, and many other things--these are questions which do not specially concern the biologist. If they are ever to receive any satisfactory explanation in terms of natural causation, this must be furnished at the hands of the psychologist. It may be possible for him to show, more satisfactorily than hitherto, that all beauty, whenever and wherever it occurs, is literally "in the eyes of the beholder"; or that objectively considered, there is no such thing as beauty. It may be--and in my opinion it probably is--purely an affair of the percipient mind itself, depending on the association of ideas with pleasure-giving objects. This association may well lead to a liking for such objects, and so to the formation of what is known as aesthetic feeling with regard to them. Moreover, beauty of inanimate nature must be an affair of the percipient mind itself, unless there be a creating intelligence with organs of sense and ideals of beauty similar to our own. And, apart from any deeper considerations, this latter possibility is scarcely entitled to be regarded as a probability, looking to the immense diversities in those ideals among different races of mankind. But, be this as it may, the scientific problem which is presented by the fact of aesthetic feeling, even if it is ever to be satisfactorily solved, is a problem which, as already remarked, must be dealt with by psychologists. As biologists we have simply to accept this feeling as a fact, and to consider how, out of such a feeling as a cause, the beauty of organic nature may have followed as an effect.

Now we have already seen how the theory of sexual selection supposes this to have happened. But against this theory a formidable objection arises, and one which I have thought it best to reserve for treatment in this place, because it serves to show the principal difference between Mr. Darwin's two great generalizations, considered as generalizations in the way of mechanical theory. For while the theory of natural selection extends equally throughout the whole range of organic nature, the theory of sexual selection has but a comparatively restricted scope, which, moreover, is but vaguely defined. For it is obvious that the theory can only apply to living organisms which are sufficiently intelligent to admit of our reasonably accrediting them with aesthetic taste--namely, in effect, the higher animals. And just as this consideration greatly restricts the possible scope of the theory, as compared with that of natural selection, so does it render undefined the zoological limits within which it can be reasonably employed. Lastly, this necessarily undefined, and yet most important limitation exposes the theory to the objection just alluded to, and which I shall now mention.

The theory, as we have just seen, is necessarily restricted in its application to the higher animals. Yet the facts which it is designed to explain are not thus restricted. For beauty is by no means restricted to the higher animals. The whole of the vegetable world, and the whole of the animal world at least as high up in the scale as the insects, must be taken as incapable of aesthetic feeling. Therefore, the extreme beauty of flowers, sea-anemones, corals, and so forth, cannot possibly be ascribed to sexual selection.

Now, with regard to this difficulty, we must begin by excluding the case of the vegetable kingdom as irrelevant. For it has been rendered highly probable--if not actually proved--by Darwin and others, that the beauty of flowers and of fruits is in large part due to natural selection. It is to the advantage of flowering plants that their organs of fructification should be rendered conspicuous--and in many cases also odoriferous,--in order to attract the insects on which the process of fertilization depends. Similarly, it is to the advantage of all plants which have brightly coloured fruits that these should be conspicuous for the purpose of attracting birds, which eat the fruits and so disseminate the seed. Hence all the gay colours and varied forms, both of flowers and fruits, have been thus adequately explained as due to natural causes, working for the welfare, as distinguished from the beauty, of the plants. For even the distribution of colours on flowers, or the beautiful patterns which so many of them present, are found to be useful in guiding insects to the organs of fructification.

Again, the green colouring of leaves, which lends so much beauty to the vegetable world, has likewise been shown to be of vital importance to the physiology of plant-life; and, therefore, may also be ascribed to natural selection. Thus, there remains only the forms of plants other than the flowers. But the forms of leaves have also in many cases been shown to be governed by principles of utility; and the same is to be said of the branching structure which is so characteristic of trees and shrubs, since this is the form most effectual for spreading out the leaves to the light and air. Here, then, we likewise find that the cause determining plant beauty is natural selection; and so we may conclude that the only reason why the forms of trees which are thus determined by utility appeal to us as beautiful, is because we are accustomed to these the most ordinary forms. Our ideas having been always, as it were, moulded upon these forms, aesthetic feeling becomes attached to them by the principle of association. At any rate, it is certain that when we contemplate almost any forms of plant-structure which, for special reasons of utility, differ widely from these (to us) more habitual forms, the result is not suggestive of beauty. Many of the tropical and un-tree-like plants--such as the cactus tribe--strike us as odd and quaint, not as beautiful. Be this however as it may, I trust I have said enough to prove that in the vegetable world, at all events, the attainment of beauty cannot be held to have been an object aimed at, so to speak, for its own sake. Even if, for the purposes of argument, we were to suppose that all the forms and colours in the vegetable world are due to special design, there could be no doubt that the purpose of this design has been in chief part a utilitarian purpose; it has not aimed at beauty exclusively for its own sake. For most of such beauty as we here perceive is plainly due to the means adopted for the attainment of life-preserving ends, which, of course, is a metaphorical way of saying that it is probably due to natural selection[50].

[50] The beauty of autumnal tints in fading leaves may possibly be adduced per contra. But here we have to remember that it is only some kinds of leaves which thus become beautiful when fading, while, even as regards those that do, it is not remarkable that their chlorophyll should, as it were, accidentally assume brilliant tints while breaking down into lower grades of chemical constitution. The case, in fact, is exactly parallel to those in the animal kingdom which are considered in the ensuing paragraphs.

Turning, then, to the animal kingdom below the level of insects, here we are bound to confess that the beauty which so often meets us cannot reasonably be ascribed either to natural or to sexual selection. Not to sexual selection for the reasons already given; the animals in question are neither sufficiently intelligent to possess any aesthetic taste, nor, as a matter of fact, do we observe that they exercise any choice in pairing. Not to natural selection, because we cannot here, as in the case of vegetables, point to any benefit as generally arising from bright colours and beautiful forms. On the principles of naturalism, therefore, we are driven to conclude that the beauty here is purely adventitious, or accidental. Nor need we be afraid to make this admission, if only we take a sufficiently wide view of the facts. For, when we do take such a view, we find that beauty here is by no means of invariable, or even of general, occurrence. There is no loveliness about an oyster or a lob-worm; parasites, as a rule, are positively ugly, and they constitute a good half of all animal species. The truth seems to be, when we look attentively at the matter, that in all cases where beauty does occur in these lower forms of animal life, its presence is owing to one of two things--either to the radiate form, or to the bright tints. Now, seeing that the radiate form is of such general occurrence among these lower animals--appearing over and over again, with the utmost insistence, even among groups widely separated from one another by the latest results of scientific classification--seeing this, it becomes impossible to doubt that the radiate form is due to some morphological reasons of wide generality. Whether these reasons be connected with the internal laws of growth, or to the external conditions of environment, I do not pretend to suggest. But I feel safe in saying that it cannot possibly be due to any design to secure beauty for its own sake. The very generality of the radiate form is in itself enough to suggest that it must have some physical, as distinguished from an aesthetic, explanation; for, if the attainment of beauty had here been the object, surely it might have been even more effectually accomplished by adopting a greater variety of typical forms--as, for instance, in the case of flowers.

Coming then, lastly, to the case of brilliant tints in the lower animals, Mr. Darwin has soundly argued that there is nothing forced or improbable in the supposition that organic compounds, presenting as they do such highly complex and such varied chemical constitutions, should often present brilliant colouring incidentally. Considered merely as colouring, there is nothing in the world more magnificent than arterial blood; yet here the colouring is of purely utilitarian significance. It is of the first importance in the chemistry of respiration; but is surely without any meaning from an aesthetic point of view. For the colour of the cheeks, and of the flesh generally, in the white races of mankind, could have been produced quite as effectually by the use of pigment--as in the case of certain monkeys. Now the fact that in the case of blood, as in that of many other highly coloured fluids and solids throughout the animal kingdom, the colour is concealed, is surely sufficient proof that the colour, if regarded from an aesthetic point of view, is accidental. Therefore, when, as in other cases, such colouring occurs upon the surface, and thus becomes apparent, are we not irresistibly led to conclude that its exhibition in such cases is likewise accidental, so far as any question of aesthetic design is concerned?

I have now briefly glanced at all the main facts of organic nature with reference to beauty; and, as a result, I think it is impossible to resist the general conclusion, that in organic nature beauty does not exist as an end per se. All cases where beauty can be pointed to in organic nature are seemingly due--either to natural selection, acting without reference to beauty, but to utility; to sexual selection, acting with reference to the taste of animals; or else to sheer accident. And if this general conclusion should be held to need any special verification, is it not to be found in the numberless cases where organic nature not only fails to be beautiful, but reveals itself as the reverse. Not again to refer to the case of parasites, what can be more unshapely than a hippopotamus, or more generally repulsive than a crocodile? If it be said that these are exceptions, and that the forms of animals as a rule are graceful, the answer--even apart from parasites--is obvious. In all cases where the habits of life are such as to render rapid locomotion a matter of utilitarian necessity, the outlines of an animal must be graceful--else, whether the locomotion be terrestrial, aerial, or aquatic, it must fail to be swift. Hence it is only in such cases as that of the hippopotamus, rhinoceros, elephant, crocodile, and so forth, where natural selection has had no concern in developing speed, that the accompanying accident of gracefulness can be allowed to disappear. But if beauty in organic nature had been in itself what may be termed an artistic object on the part of a divine Creator, it is absurd to suggest that his design in this matter should only have been allowed to appear where we are able to detect other and very good reasons for its appearance.

* * * * *

Thus, whether we look to the facts of adaptation or to those of beauty, everywhere throughout organic nature we meet with abundant evidence of natural causation, while nowhere do we meet with any independent evidence of supernatural design. But, having led up to this conclusion, and having thus stated it as honestly as I can, I should like to finish by further stating what, in my opinion is its logical bearing upon the more fundamental tenets of religious thought.

As I have already observed at the commencement of this brief exposition, prior to the Darwinian theory of organic evolution, the theologian was prone to point to the realm of organic nature as furnishing a peculiarly rich and virtually endless store of facts, all combining in their testimony to the wisdom and the beneficence of the Deity. Innumerable adaptations of structures to functions appeared to yield convincing evidence in favour of design; the beauty so profusely shed by living forms appeared to yield evidence, no less convincing, of that design as beneficent. But both these sources of evidence have now, as it were, been tapped at their fountain-head: the adaptation and the beauty are alike receiving their explanation at the hands of a purely mechanical philosophy. Nay, even the personality of man himself is assailed; and this not only in the features which he shares with the lower animals, but also in his god-like attributes of reason, thought, and conscience. All nature has thus been transformed before the view of the present generation in a manner and to an extent that has never before been possible: and inasmuch as the change which has taken place has taken place in the direction of naturalism, and this to the extent of rendering the mechanical interpretation of nature universal, it is no wonder if the religious mind has suddenly awakened to a new and a terrible force in the words of its traditional enemy-- Where is now thy God?

This is not the place to discuss the bearings of science on religion[51]; but I think it is a place where one may properly point out the limits within which no such bearings obtain. Now, from what has just been said, it will be apparent that I am not going to minimise the change which has been wrought. On the contrary, I believe it is only stupidity or affectation which can deny that the change in question is more deep and broad than any single previous change in the whole history of human thought. It is a fundamental, a cosmical, a world-transforming change. Nevertheless, in my opinion, it is a change of a non-theistic, as distinguished from an a-theistic, kind. It has rendered impossible the appearance in literature of any future Paley, Bell, or Chalmers; but it has done nothing in the way of negativing that belief in a Supreme Being which it was the object of these authors to substantiate. If it has demonstrated the futility of their proof, it has furnished nothing in the way of disproof. It has shown, indeed, that their line of argument was misjudged when they thus sought to separate organic nature from inorganic as a theatre for the special or peculiar display of supernatural design; but further than this it has not shown anything. The change in question therefore, although greater in degree, is the same in kind as all its predecessors: like all previous advances in cosmological theory which have been wrought by the advance of science, this latest and greatest advance has been that of revealing the constitution of nature, or the method of causation, as everywhere the same. But it is evident that this change, vast and to all appearance final though it be, must end within the limits of natural causation itself. The whole world of life and mind may now have been annexed to that of matter and energy as together constituting one magnificent dominion, which is everywhere subject to the same rule, or method of government. But the ulterior and ultimate question touching the nature of this government as mental or non-mental, personal or impersonal, remains exactly where it was. Indeed, this is a question which cannot be affected by any advance of science, further than science has proved herself able to dispose of erroneous arguments based upon ignorance of nature. For while the sphere of science is necessarily restricted to that of natural causation which it is her office to explore, the question touching the nature of this natural causation is one which as necessarily lies without the whole sphere of such causation itself: therefore it lies beyond any possible intrusion by science. And not only so. But if the nature of natural causation be that of the highest order of known existence, then, although we must evidently be incapable of conceiving what such a Mind is, at least we seem capable of judging what in many respects it is not. It cannot be more than one; it cannot be limited either in space or time; it cannot be other than at least as self-consistent as its manifestations in nature are invariable. Now, from the latter deduction there arises a point of first-rate importance in the present connexion. For if the so-called First Cause be intelligent, and therefore all secondary causes but the expression of a supreme Will, in as far as such a Will is self-consistent, the operation of all natural causes must be uniform,--with the result that, as seen by us, this operation must needs appear to be what we call mechanical. The more unvarying the Will, the more unvarying must be this expression thereof; so that, if the former be absolutely self-consistent, the latter cannot fail to be as reasonably interpreted by the theory of mindless necessity, as by that of ubiquitous intention. Such being, as it appears to me, the pure logic of the matter, the proof of organic evolution amounts to nothing more than the proof of a natural process. What mode of being is ultimately concerned in this process--or in what it is that this process ultimately consists--is a question upon which science is as voiceless as speculation is vociferous.

[51] The best treatise on this subject is Prof. Le Conte's Evolution and its Relation to Religious Thought (Appleton & Co. 1888).

But, it may still be urged, surely the principle of natural selection (with its terrible basis in the struggle for existence) and the principle of sexual selection (with its consequence in denying beauty to be an end in itself) demonstrate that, if there be design in nature, such design at all events cannot be beneficent. To this, however, I should again reply that, just as touching the major question of design itself, so as touching this minor question of the quality of such design as beneficent, I do not see how the matter has been much affected by a discovery of the principles before us. For we did not need a Darwin to tell us that the whole creation groaneth and travaileth together in pain. The most that in this connexion Darwin can fairly be said to have done is to have estimated in a more careful and precise manner than any of his predecessors, the range and the severity of this travail. And if it be true that the result of what may be called his scientific analysis of nature in respect of suffering is to have shown the law of suffering even more severe, more ubiquitous, and more necessary than it had ever been shown before, we must remember at the same time how he has proved, more rigidly than was ever proved before, that suffering is a condition to improvement--struggle for life being the raison d'être of higher life, and this not only in the physical sphere, but also in the mental and moral.

Lastly, if it be said that the choice of such a method, whereby improvement is only secured at the cost of suffering, indicates a kind of callousness on the part of an intelligent Being supposed to be omnipotent, I confess that such does appear to me a legitimate conclusion--subject, however, to the reservation that higher knowledge might displace it. For, as far as matters are now actually presented to the unbiased contemplation of a human mind, this provisional inference appears to me unavoidable--namely, that if the world of sentient life be due to an Omnipotent Designer, the aim or motive of the design must have been that of securing a continuous advance of animal improvement, without any regard at all to animal suffering. For I own it does not seem to me compatible with a fair and honest exercise of our reason to set the sum of animal happiness over against the sum of animal misery, and then to allege that, in so far as the former tends to balance--or to over-balance--the latter, thus far is the moral character of the design as a whole vindicated. Even if it could be shown that the sum of happiness in the brute creation considerably preponderates over that of unhappiness--which is the customary argument of theistic apologists,--we should still remain without evidence as to this state of matters having formed any essential part of the design. On the other hand, we should still be in possession of seemingly good evidence to the contrary. For it is clearly a condition to progress by survival of the fittest, that as soon as organisms become sentient selection must be exercised with reference to sentiency; and this means that, if further progress is to take place, states of sentiency must become so organized with reference to habitual experience of the race, that pleasures and pains shall answer respectively to states of agreement and disagreement with the sentient creature's environment. Those animals which found pleasure in what was deleterious to life would not survive, while those which found pleasure in what was beneficial to life would survive; and so eventually, in every species of animal, states of sentiency as agreeable or disagreeable must approximately correspond with what is good for the species or bad for the species. Indeed, we may legitimately surmise that the reason why sentiency (and, a fortiori, conscious volition) has ever appeared upon the scene at all, has been because it furnishes--through this continuously selected adjustment of states of sentiency to states of the sentient organism--so admirable a means of securing rapid, and often refined, adjustments by the organism to the habitual conditions of its life[52]. But, if so, not only is this state of matters a condition to progress in the future; it is further, and equally, a consequence of progress in the past.

[52] See Mental Evolution in Animals, pp. 110-111.

However, be this as it may, from all that has gone before does it not become apparent that pleasure or happiness on the one hand, and pain or misery on the other, must be present in sentient nature? And so long as they are both seen to be equally necessary under the process of evolution by natural selection, we have clearly no more reason to regard the pleasure than the pain as an object of the supposed design. Rather must we see in both one and the same condition to progress under the method of natural causation which is before us; and therefore I cannot perceive that it makes much difference--so far as the argument for beneficence is concerned--whether the pleasures of animals outweigh their pains, or vice vers?

Upon the whole, then, it seems to me that such evidence as we have is against rather than in favour of the inference, that if design be operative in animate nature it has reference to animal enjoyment or well-being, as distinguished from animal improvement or evolution. And if this result should be found distasteful to the religious mind--if it be felt that there is no desire to save the evidences of design unless they serve at the same time to testify to the nature of that design as beneficent,--I must once more observe that the difficulty thus presented to theism is not a difficulty of modern creation. On the contrary, it has always constituted the fundamental difficulty with which natural theologians have had to contend. The external world appears, in this respect, to be at variance with our moral sense; and when the antagonism is brought home to the religious mind, it must ever be with a shock of terrified surprise. It has been newly brought home to us by the generalizations of Darwin; and therefore, as I said at the beginning, the religious thought of our generation has been more than ever staggered by the question--Where is now thy God? But I have endeavoured to show that the logical standing of the case has not been materially changed; and when this cry of Reason pierces the heart of Faith, it remains for Faith to answer now, as she has always answered before--and answered with that trust which is at once her beauty and her life--Verily thou art a God that hidest thyself.

APPENDIX AND NOTES

APPENDIX TO CHAPTER V.

ON OBJECTIONS WHICH HAVE BEEN BROUGHT AGAINST THE THEORY OF ORGANIC EVOLUTION ON GROUNDS OF PAL 芰 NTOLOGY.

While stating in the text, and in a necessarily general way, the evidence which is yielded by Paleontology to the theory of organic evolution, I have been desirous of not overstating it. Therefore, in the earlier paragraphs of the chapter, which deal with the most general heads of such evidence, I introduced certain qualifying phrases; and I will now give the reasons which led me to do so.

Of all the five biological sciences which have been called into evidence--viz. those of Classification, Morphology, Embryology, Paleontology, and Geographical Distribution--it is in the case of Paleontology alone that any important or professional opinions still continue to be unsatisfied. Therefore, in order that justice may be done to this line of dissent, I have thought it better to deal with the matter in a separate Appendix, rather than to hurry it over in the text. And, as all the difficulties or objections which have been advanced against the theory of evolution on grounds of Paleontology must vary, as to their strength, with the estimate which is taken touching the degree of imperfection of the geological record, I will begin by adding a few paragraphs to what has already been said in the text upon this subject.

First, then, as to the difficulties in the way of fossils being formed at all. We have already noticed in the text that it is only the more or less hard parts of organisms which under any circumstances can be fossilized; and even the hardest parts quickly disintegrate if not protected from the weather on land, or from the water on the sea-bottom. Moreover, as Darwin says, "we probably take a quite erroneous view when we assume that sediment is being deposited over nearly the whole bed of the sea, at a rate sufficiently quick to embed and preserve fossil remains. Throughout an enormously large proportion of the ocean, the bright blue tint of the water bespeaks its purity. The many cases on record of a formation conformably covered, after an immense interval of time, by another and a later formation, without the underlying bed having suffered in the interval any wear and tear, seem explicable only on the view of the bottom of the sea not rarely lying for ages in an unaltered condition." Next, as regards littoral animals, he shows the difficulty which they must have in becoming fossils, and gives a striking example in several of the existing species of a sub-family of cirripedes (Chthamalin?, "which coat the rocks all over the world in infinite numbers," yet, with the exception of one species which inhabits deep water, no vestige of any of them has been found in any tertiary formation, although it is known that the genus Chthamalus existed through the Chalk period. Lastly, "with respect to the terrestrial productions which lived through the secondary and paleozoic periods, it is superfluous to state our evidence is fragmentary in an extreme degree. For instance, until recently not a land shell was known belonging to either of these vast periods," with one exception; while, "in regard to mammiferous remains, a glance at the historical table in Lyell's Manual will bring home the truth, how accidental and rare has been their preservation, far better than pages of detail. Nor is their rarity surprising, when we remember how large a proportion of the bones of tertiary mammals have been discovered either in caves or in lacustrine deposits; and that not a cave or true lacustrine bed is known belonging to the age of our secondary or paleozoic formations."

But perhaps of even more importance than all these known causes which prevent the formation of fossils, is the existence of unknown causes which make for the same result. For example, the Flysch-formation is a formation of several thousand feet in thickness (as much as 6000 in some places), and it extends for at least 300 miles from Vienna to Switzerland; moreover, it consists of shale and sandstone. Therefore, alike in respect of time, space, and character, it is just such a formation as we should expect to find highly rich in fossils; yet, "although this great mass has been most carefully searched, no fossils, except a few vegetable remains, have been found."

So much then for the difficulty, so to speak, which nature experiences in the manufacture of fossils. Probably not one per cent. of the species of animals which have inhabited the earth has left a single individual as a fossil, whereby to record its past existence.

But of even more importance than this difficulty of making fossils in the first instance, is the difficulty of preserving them when they are made. The vast majority of fossils have been formed under water, and a large proportional number of these--whether the animals were marine, terrestrial, or inhabitants of fresh water--have been formed in sedimentary deposits either of sand, gravel, or other porous material. Now, where such deposits have been afterwards raised into the air for any considerable time--and this has been more or less the case with all deposits which are available for exploration--their fossiliferous contents will have been, as a general rule, dissolved by the percolation of rain-water charged with carbonic acid. Similarly, sea-water has recently been found to be a surprisingly strong solvent of calcareous material: hence, Saturn-like, the ocean devours her own progeny as far as shells and bones of all kinds are concerned--and this to an extent of which we have probably no adequate conception.

Of still greater destructive influence, however, than these solvent agencies in earth and sea, are the erosive agencies of both. Any one who watches the pounding of the waves upon the shore; who then observes the effect of it upon the rocks broken into shingle, and on the shingle reduced to sand; who, looking behind him at the cliffs, sees there the evidence of the gradual advance of this all-pulverising power--an advance so gradual that no yard of it is accomplished until within that yard the "white teeth" have eaten well into the "bowels of the earth"; who then reflects that this process is going on simultaneously over hundreds of thousands of miles of coast-lines throughout the world; and who finally extends his mental vision from space to time, by trying dimly to imagine what this ever-roaring monster must have consumed during the hundreds of millions of years that slowly rising and slowly sinking continents have exposed their whole areas to her jaws; whoever thus observes and thus reflects must be a dull man, if he does not begin to feel that in the presence of such a destroyer as this we have no reason to wonder at a frequent silence in the testimony of the rocks.

But although the erosive agency of the sea is thus so inconceivably great, it is positively small if compared with erosive agencies on land. The constant action of rain, wind, and running water, in wearing down the surfaces of all lands into "the dust of continents to be"; the disintegrating effects on all but the very hardest rocks of winter frosts alternating with summer heats; the grinding power of ice in periods of glaciation; and last, but not least, the wholesale melting up of sedimentary formations whenever these have sunk for any considerable distance beneath the earth's surface:--all these agencies taken together constitute so prodigious a sum of energies combined through immeasureable ages in their common work of destruction, that when we try to realise what it must amount to, we can scarcely fail to wonder, not that the geological record is highly imperfect, but that so much of the record has survived as we find to have been the case. And, if we add to these erosive and solvent agencies on land the erosive and solvent agencies of the sea, we may almost begin to wonder that anything deserving the name of a geological record is in existence at all.

That such estimates of the destructive powers of nature are not mere matters of speculative reasoning may be amply shown by stating one single fact, which, like so many others where the present subject is concerned, we owe to the generalizations of Darwin. Plutonic rocks, being those which have emerged from subterranean heat of melting intensity, must clearly at some time or another have lain beneath the whole thickness of sedimentary deposits, which at that time occupied any part of the earth's surface where we now find the Plutonic rocks exposed to view. Or, in other words, wherever we now find Plutonic rocks at the surface of the earth, we must conclude that all the sedimentary rocks by which they were covered when in a molten state have since been entirely destroyed; several vertical miles of the only kinds of rocks in which fossils can possibly occur must in all such cases have been abolished in toto. Now, in many parts of the world metamorphic rocks--which have thus gradually risen from Plutonic depths, while miles of various other rock-formations have been removed from their now exposed surfaces--cover immense areas, and therefore testify by their present horizontal range, no less than by their previously vertical depth, to the enormous scale on which a total destruction has taken place of everything that once lay above them. For instance, the granitic region of Parime is at least nineteen times the size of Switzerland; a similar region south of the Amazon is probably larger than France, Spain, Italy, and Great Britain all put together; and, more remarkable still, over the area of the United States and Canada, granitic rocks exceed in the proportion of 19 to 12-1/2 the whole of the newer Paleozoic formations. Lastly, after giving these examples, Darwin adds the important consideration, that "in many regions the metamorphic and granitic rocks would be found much more widely extended than they appear to be, if all the sedimentary beds were removed which rest unconformably on them, and which could not have formed part of the original mantle under which they were crystallized."

The above is a brief condensation of the already condensed statement which Darwin has given of the imperfection of the geological record; but I think it is enough to show, in a general way, how precarious must be the nature of any objections to the theory of evolution which are founded merely upon the silence of Paleontology in cases where, if the record were anything like complete, we should be entitled to expect from it some positive information. But, as we have seen in the text, imperfect though the record be, in as far as it furnishes positive information at all, this is well-nigh uniformly in favour of the theory; and therefore, even on grounds of Paleontology alone, it appears to me that Darwin is much too liberal where he concludes his discussion by saying,--"Those who believe that the geological record is in any degree perfect, will undoubtedly at once reject the theory." If in any measure reasonable, such persons ought rather to examine their title to such a belief; and even if they disregard the consensus of testimony which is yielded by all the biological sciences to the theory of evolution, they ought at least to hold their judgment in suspense until they shall have not only set against the apparently negative testimony which is yielded by geology its unquestionably positive testimony, but also well considered the causes which may--or rather must--have so gravely impaired the geological record.

However, be this as it may, I will now pass on to consider the difficulties and objections which have been brought against the theory on grounds of Paleontology.

These may be classified under four heads. First, the absence of varietal links between allied species; second, the sudden appearance of whole groups of species--not only as genera and families, but even sometimes as orders and classes--without any forms leading up to them; third, the occurrence of highly organized types at much lower levels of geological strata than an evolutionist would antecedently expect; and, fourth, the absence of fossils of any kind lower down than the Cambrian strata.

Now all these objections depend on estimates of the imperfection of the geological record much lower than that which is formed by Darwin. Therefore I have arranged the objections in their order of difficulty in this respect, or in the order that requires successively increasing estimates of the imperfection of the record, if they are to be successively answered.

I think that the first of them has been already answered in the text, by showing that even a very moderate estimate of the imperfection of the record is enough to explain why intermediate varieties, connecting allied species, are but comparatively seldom met with. Moreover it was shown that in some cases, where shells are concerned, remarkably well-connected series of such varieties have been met with. And the same applies to species and genera in certain other cases, as in the equine family.

But no doubt a greater difficulty arises where whole groups of species and genera, or even families and orders, appear to arise suddenly, without anything leading up to them. Even this the second difficulty, however, admits of being fully met, when we remember that in very many cases it has been proved, quite apart from the theory of descent, that superjacent formations have been separated from one another by wide intervals of time. And even although it often happens that intermediate deposits which are absent in one part of the world are present in another, we have no right to assume that such is always the case. Besides, even if it were, we should have no right further to assume that the faunas of widely separated geographical areas were identical during the time represented by the intermediate formation. Yet, unless they were identical, we should not expect the fossils of the intermediate formation, where extant, to yield evidence of what the fossils would have been in this same formation elsewhere, had it not been there destroyed. Now, as a matter of fact, "geological formations of each region are almost invariably intermittent"; and although in many cases a more or less continuous record of past forms of life can be obtained by comparing the fossils of one region and formation with those of another region and adjacent formations, it is evident (from what we know of the present geographical distribution of plants and animals) that not a few cases there must have been where the interruption of the record in one region cannot be made good by thus interpolating the fossils of another region. And we must remember it is by selecting the cases where this cannot be done that the objection before us is made to appear formidable. In other words, unless whole groups of new species which are unknown in formation A appear suddenly in formation C of one region (X), where the intermediate formation B is absent; and unless in some other region (Y), where B is present, the fossiliferous contents of B fail to supply the fossil ancestry of the new species in A (X); unless such a state of matters is found to obtain, the objection before us has nothing to say. But at best this is negative evidence; and, in order to consider it fairly, we ought to set against it the cases where an interposition of fossils found in B (Y) does furnish the fossil ancestry of what would otherwise have been an abrupt appearance of whole groups of new species in A (X). Now such cases are neither few nor unimportant, and therefore they deprive the objection of the force it would have had if the selected cases to the contrary were the general rule.

In addition to these considerations, the following, some of which are of a more special kind, appear to me so important that I will quote them almost in extenso.

We continually forget how large the world is, compared with the area over which our geological formations have been carefully examined: we forget that groups of species may elsewhere have long existed, and have slowly multiplied, before they invaded the ancient archipelagoes of Europe and the United States. We do not make due allowance for the intervals of time which have elapsed between our consecutive formations,--longer perhaps in many cases than the time required for the accumulation of each formation. These intervals will have given time for the multiplication of species from some one parent form; and, in the succeeding formation, such groups of species will appear as if suddenly created.

I may here recall a remark formerly made, namely, that it might require a long succession of ages, to adapt an organism to some new and peculiar line of life, for instance, to fly through the air; and consequently that the transitional form would often long remain confined to some one region; but that, when this adaptation had once been effected, and a few species had thus acquired a great advantage over other organisms, a comparatively short time would be necessary to produce many divergent forms, which would spread rapidly and widely throughout the world....

In geological treatises, published not many years ago, mammals were always spoken of as having abruptly come in at the commencement of the tertiary series. And now one of the richest known accumulations of fossil mammals belongs to the middle of the secondary series; and true mammals have been discovered in the new red sandstone at nearly the commencement of this great series. Cuvier used to urge that no monkey occurred in any tertiary stratum; but now extinct species have been discovered in India, South America, and in Europe as far back as the miocene stage. Had it not been for the rare accident of the preservation of footsteps in the new red sandstone of the United States, who would have ventured to suppose that, no less than at least thirty kinds of bird-like animals, some of gigantic size, existed during that period? Not a fragment of bone has been discovered in these beds. Not long ago paleontologists maintained that the whole class of birds came suddenly into existence during the eocene period; but now we know, on the authority of Professor Owen, that a bird certainly lived during the deposition of the upper green-sand. And still more recently that strange bird, the Archeopteryx ... has been discovered in the oolitic slates of Solenhofen. Hardly any recent discovery shows more forcibly than this, how little we as yet know of the former inhabitants of the world.

I may give another instance, which, from having passed under my own eyes, has much struck me. In a memoir on Fossil Sessile Cirripedes, I stated that, from the number of existing and extinct tertiary species; from the extraordinary abundance of the individuals of many species all over the world from the Arctic regions to the equator, inhabiting various zones of depths from the upper tidal limits to 50 fathoms; from the perfect manner in which specimens are preserved in the oldest tertiary beds; from the ease with which even a fragment of a valve can be recognized; from all these circumstances, I inferred that had sessile cirripedes existed during the secondary periods, they would certainly have been preserved and discovered; and as not one species had then been discovered in beds of this age, I concluded that this great group had been suddenly developed at the commencement of the tertiary series. This was a sore trouble to me, adding as I thought one more instance of the abrupt appearance of a great group of species. But my work had hardly been published, when a skilful paleontologist, M. Bosquet, sent me a drawing of a perfect specimen of an unmistakeable sessile cirripede, which he had himself extracted from the chalk of Belgium. And, as if to make the case as striking as possible, this sessile cirripede was a Chthamalus, a very common, large, and ubiquitous genus, of which not one specimen has as yet been found even in any tertiary stratum. Still more recently, a Pyrgoma, a member of a distinct sub-family of sessile cirripedes, has been discovered by Mr. Woodward in the upper chalk; so that we now have abundant evidence of the existence of this group of animals during the secondary period.

The case most frequently insisted on by paleontologists of the apparently sudden appearance of a whole group of species, is that of the teleostean fishes, low down, according to Agassiz, in the Chalk period. This group includes the large majority of existing species. But certain Jurassic and Triassic forms are now commonly admitted to be teleostean; and even some paleozoic forms have been thus classed by one high authority. If the teleosteans had really appeared suddenly in the northern hemisphere, the fact would have been highly remarkable; but it would not have formed an insuperable difficulty, unless it could likewise have been shown that at the same period the species were suddenly and simultaneously developed in other quarters of the world. It is almost superfluous to remark that hardly any fossil fish are known from south of the equator; and by running through Pictet's Paleontology it will be seen that very few species are known from several formations in Europe. Some few families of fish now have a confined range; the teleostean fish might formerly have had a similarly confined range, and after having been largely developed in some one sea, might have spread widely. Nor have we any right to suppose that the seas of the world have always been so freely open from south to north as they are at present. Even at this day, if the Malay Archipelago were converted into land, the tropical parts of the Indian Ocean would form a large and perfectly enclosed basin, in which any great group of marine animals might be multiplied; and here they would remain confined, until some of the species became adapted to a cooler climate, and were enabled to double the southern capes of Africa or Australia, and thus reach other and distant seas.

From these considerations, from our ignorance of the geology of other countries beyond the confines of Europe and the United States; and from the revolution in our paleontological knowledge effected by the discoveries of the last dozen years, it seems to me to be about as rash to dogmatize on the succession of organic forms throughout the world, as it would be for a naturalist to land for five minutes on some one barren point in Australia, and then to discuss the number and range of its productions[53].

[53] Origin of Species, 282-5.

In view of all the foregoing facts and considerations, it appears to me that the second difficulty on our list is completely answered. Indeed, even on a moderate estimate of the imperfection of the geological record, the wonder would have been if many cases had not occurred where groups of species present the fictitious appearance of having been suddenly and simultaneously created in the particular formations where their remains now happen to be observable.

Turning next to the third objection, there cannot be any question that every here and there in the geological series animals occur of a much higher grade zoologically than the theory of evolution would have expected to find in the strata where they are found. At any rate, speaking for myself, I should not have antecedently expected to meet with such highly differentiated insects as butterflies and dragonflies in the middle of the Secondaries: still less should I have expected to encounter beetles, cockroaches, spiders, and May-flies in the upper and middle Primaries--not to mention an insect and a scorpion even in the lower. And I think the same remark applies to a whole sub-kingdom in the case of Vertebrata. For although it is only the lowest class of the sub-kingdom which, so far as we positively know, was represented in the Devonian and Silurian formations, we must remember, on the one hand, that even a cartilaginous or ganoid fish belongs to the highest sub-kingdom of the animal series; and, on the other hand, that such animals are thus proved to have abounded in the very lowest strata where there is good evidence of there having been any forms of life at all. Lastly, the fact that Marsupials occur in the Trias, coupled with the fact that the still existing Monotremata are what may be termed animated fossils, referring us by their lowly type of organization to some period enormously more remote,--these facts render it practically certain that some members of this very highest class of the highest sub-kingdom must have existed far back in the Primaries.

These things, I say, I should not have expected to find, and I think all other evolutionists ought to be prepared to make the same acknowledgment. But as these things have been found, the only possible way of accounting for them on evolutionary principles is by supposing that the geological record is even more imperfect than we needed to suppose in order to meet the previous objections. I cannot see, however, why evolutionists should be afraid to make this acknowledgment. For I do not know any reason which would lead us to suppose that there is any common measure between the distances marked on our tables of geological formations, and the times which those distances severally represent. Let the reader turn to the table on page 163, and then let him say why the 30,000 feet of so-called Azoic rocks may not represent a greater duration of time than does the thickness of all the Primary rocks above them put together. For my own part I believe that this is probably the case, looking to the enormous ages during which these very early formations must have been exposed to destructive agencies of all kinds, now at one time and now at another, in different parts of the world. And, of course, we are without any means of surmising what ranges of time are represented by the so-called Primeval rocks, for the simple reason that they are non-sedimentary, and non-sedimentary rocks cannot be expected to contain fossils.

But, it will be answered, the 30,000 feet of Azoic rocks, lying above the Primeval, are sedimentary to some extent: they are not all completely metamorphic: yet they are all destitute of fossils. This is the fourth and last difficulty which has to be met, and it can only be met by the considerations which have been advanced by Lyell and Darwin. The former says:--

The total absence of any trace of fossils has inclined many geologists to attribute the origin of the most ancient strata to an azoic period, or one antecedent to the existence of organic beings. Admitting, they say, the obliteration, in some cases, of fossils by plutonic action, we might still expect that traces of them would oftener be found in certain ancient systems of slate, which can scarcely be said to have assumed a crystalline structure. But in urging this argument it seems to be forgotten that there are stratified formations of enormous thickness, and of various ages, some of them even of tertiary date, and which we know were formed after the earth had become the abode of living creatures, which are, nevertheless, in some districts, entirely destitute of all vestiges of organic bodies[54].

[54] Elements of Geology, p. 587.

He then proceeds to mention sundry causes (in addition to plutonic action) which are adequate to destroy the fossiliferous contents of stratified rocks, and to show that these may well have produced enormous destruction of organic remains in these oldest of known formations.

Darwin's view is that, during the vast ages of time now under consideration, it is probable that the distribution of sea and land over the earth's surface has not been uniformly the same, even as regards oceans and continents. Now, if this were the case, "it might well happen that strata which had subsided some miles nearer to the centre of the earth, and which had been pressed on by an enormous weight of superincumbent water, might have undergone far more metamorphic action than strata which have always remained nearer to the surface. The immense areas in some parts of the world, for instance in South America, of naked metamorphic rocks, which must have been heated under great pressure, have always seemed to me to require some special explanation; and we may perhaps believe that we see, in these large areas, the many formations long anterior to the Cambrian epoch in a completely metamorphosed and denuded condition[55]." The probability of this view he sustains by certain general considerations, as well as particular facts touching the geology of oceanic islands, &c.

[55] Origin of Species, p. 289.

On the whole, then, it seems to me but reasonable to conclude, with regard to all four objections in question, as Darwin concludes with regard to them:--

For my part, following out Lyell's metaphor, I look at the geological record as a history of the world imperfectly kept, written in a changing dialect; of this history we possess the last volume alone, relating only to two or three countries. Of this volume, only here and there a short chapter has been preserved; and of each page only here and there a few lines. Each word of the slowly-changing language, more or less different in the successive chapters, may represent the forms of life, which are entombed in our consecutive formations, and which falsely appear to us to have been abruptly introduced. On this view, the difficulties above discussed are greatly diminished, or even disappear[56].

[56] Ibid.

As far as I can see, the only reasonable exception that can be taken to this general view of the whole matter, is one which has been taken from the side of astronomical physics.

Put briefly, it is alleged by one of the highest authorities in this branch of science, that there cannot have been any such enormous reaches of unrecorded time as would be implied by the supposition of there having been a lost history of organic evolution before the Cambrian period. The grounds of this allegation I am not qualified to examine; but in a general way I agree with Prof. Huxley in feeling that, from the very nature of the case, they are necessarily precarious,--and this in so high a degree that any conclusions raised on such premises are not entitled to be deemed formidable[57].

[57] See Lay Sermons, Lecture on Geological Reform.

* * * * *

Turning now to plants, the principal and the ablest opponent of the theory of evolution is here unquestionably Mr. Carruthers[58]. The difficulties which he adduces may be classified under three heads, as follows:--

[58] See especially the following Presidential addresses:--Geol. Assoc. Nov. 1876; Section D. Brit. Assoc., 1886; Lin. Soc., 1890.

1. There is no evidence of change in specific forms of existing plants. Not only are the numerous species of plants which have been found in Egyptian mummies indistinguishable from their successors of to-day; but, what is of far more importance, a large number of our own indigenous plants grew in Great Britain during the glacial period (including under this term the warm periods

between those of successive glaciations), and in no one case does it appear that any modification of specific type has occurred. This fact is particularly remarkable as regards leaves, because on the one hand they are the organs of plants which are most prone to vary, while on the other hand they are likewise the organs which lend themselves most perfectly to the process of fossilization, so that all details of their structure can be minutely observed in the fossil state. Yet the interval since the glacial period, although not a long one geologically speaking, is certainly what may be called an appreciable portion of time in the history of Dicotyledonous plants since their first appearance in the Cretaceous epoch. Again, if we extend this kind of enquiry so as to include the world as a whole, a number of other species of plants dating from the glacial epoch are found to tell the same story-- notwithstanding that, in the opinion of Mr. Carruthers, they must all have undergone many changes of environment while advancing before, and retreating after, successive glaciations in different parts of the globe. Or, to quote his own words:--"The various physical conditions which of necessity affected these {41} species in their diffusion over such large areas of the earth's surface in the course of, say, 250,000 years, should have led to the production of many varieties; but the uniform testimony of the remains of this considerable pre-glacial flora, as far as the materials admit of a comparison, is that no appreciable change has taken place."

2. There is no appearance of generalized forms among the earliest plants with which we are acquainted. For example, in the first dry land flora--the Devonian--we have representatives of the Filices, Equisetace? and Lycopodiace? all as highly specialized as their living representatives, and exhibiting the differential characters of these closely related groups. Moreover, these plants were even more highly organized than their existing descendants in regard to their vegetative structure, and in some cases also in regard to their reproductive organs. So likewise the Gymnosperms of that time show in their fossil state the same highly organized woody structure as their living representatives.

3. Similarly, and more generally, the Dicotyledonous plants, which first appear in the Cretaceous rocks, appear there suddenly, without any forms leading up to them--notwithstanding that "we know very well the extensive flora of the underlying Wealden." Moreover, we have all the three great divisions of the Dicotyledons appearing together, and so highly differentiated

that all the species are referred to existing genera, with the exception of a very few imperfectly preserved, and therefore uncertain fragments.

Such being the facts, we may begin by noticing that, even at first sight, they present different degrees of difficulty. Thus, I cannot see that there is much difficulty with regard to those in class 2. Only if we were to take the popular (and very erroneous) view of organic evolution as a process which is always and everywhere bound to promote the specialization of organic types--only then ought we to see any real difficulty in the absence of generalized types preceding these existing types. Of course we may wonder why still lower down in the geological series we do not meet with more generalized (or ancestral) types; but this is the difficulty number 3, which we now proceed to examine.

Concerning the other two difficulties, then, the only possible way of meeting that as to the absence of any parent forms lower down in the geological series is by falling back--as in the analogous case of animals--upon the imperfection of the geological record. Although it is certainly remarkable that we should not encounter any forms serving to connect the Dicotyledonous plants of the Chalk with the lower forms of the underlying Wealden, we must again remember that difficulties thus depending on the absence of any corroborative record, are by no means equivalent to what would have arisen in the presence of an adverse record--such, for instance, as would have been exhibited had the floras of the Wealden and the Chalk been inverted. But, as the case actually stands, the mere fact that Dicotyledonous plants, where they first occur, are found to have been already differentiated into their three main divisions, is in itself sufficient evidence, on the general theory of evolution, that there must be a break in the record as hitherto known between the Wealden and the Chalk. Nor is it easy to see how the opponents of this theory can prove their negative by furnishing evidence to the contrary. And although such might justly be deemed an unfair way of putting the matter, were this the only case where the geological record is in evidence, it is not so when we remember that there are numberless other cases where the geological record does testify to connecting links in a most satisfactory manner. For in view of this consideration the burden of proof is thrown upon those who point to particular cases where there is thus a conspicuous absence of transitional forms--the burden, namely, of proving that such cases are not due merely to a break in the record. Besides, the break in the record as regards this particular case may be apparent rather than real. For I suppose there is no greater authority on the pure geology of the subject than Sir Charles Lyell, and this is what he says of the particular case in question. "If the passage seem at present to be somewhat sudden from the flora of the Lower or Neocomian to that of the Upper Cretaceous period, the abruptness of the change will probably disappear when we are better acquainted with the fossil vegetation of the uppermost tracts of the Neocomian and that of the lowest strata of the Gault, or true Cretaceous series[59]."

[59] Elements of Geology, p. 280.

Lastly, the fact of the flora of the glacial epoch not having exhibited any modifications during the long residence of some of its specific types in Great Britain and elsewhere, is a fact of some importance to the general theory of evolution, since it shows a higher degree of stability on the part of these specific types than might perhaps have been expected, supposing the theory to be true. But I do not see that this constitutes a difficulty against the theory, when we have so many other cases of proved transmutation to set against it. For instance, not to go further afield than this very glacial flora itself, it will be remembered that in an earlier chapter I selected it as furnishing specially cogent proof of the transmutation of species. What, then, is the explanation of so extraordinary a difference between Mr. Carruthers' views and my own upon this point? I believe the explanation to be that he does not take a sufficiently wide survey of the facts.

To begin with, it seems to me that he exaggerates the vicissitudes to which the species of plants that he calls into evidence have been exposed while advancing before, and retreating after, the ice. Rather do I agree with Darwin that "they would not have been exposed during their long migrations to any great diversity of temperature; and as they all migrated in a body together, their mutual relations will not have been much disturbed; hence, in accordance with the principles indicated in this volume, these forms will not have been liable to much modification[60]." But, be this matter of opinion as it may, a much better test is afforded by those numerous cases all the world over, where arctic species have been left stranded on alpine areas by the retreat of glaciation; because here there is no room for differences of opinion as to a "change of environment" having taken place. Not to speak of climatic differences between arctic and alpine stations, consider merely the changes which must have taken place in the relations of the thus isolated species to each other, as well as to those of all the foreign plants, insects, &c., with which they have long been thrown into close association. If in such cases no variation or transmutation had taken place since the glacial epoch, then indeed there would have been a difficulty of some magnitude. But, by parity of reasoning, whatever degree of difficulty would have been thus presented is not merely discharged, but converted into at least an equal degree of corroboration, when it is found that under such circumstances, in whatever part of the world they have occurred, some considerable amount of variation and transmutation has always taken place,--and this in the animals as well as in the plants. For instance, again to quote Darwin, "If we compare the present Alpine plants and animals of the several great European mountain-ranges one with another, though many of the species remain identically the same, some exist as varieties, some as doubtful forms or sub-species, and some as distinct yet closely allied species representing each other on the several ranges[61]." Lastly, if instead of considering the case of alpine floras, we take the much larger case of the Old and New World as a whole, we meet with much larger proofs of the same general facts. For, "during the slowly decreasing warmth of the Pliocene period, as soon as the species in common, which inhabited the New and Old Worlds, migrated south of the Polar Circle, they will have been completely cut off from each other. This separation, as far as the more temperate productions are concerned, must have taken place long ages ago. As the plants and animals migrated southward, they will have become mingled in one great region with the native American productions, and would have had to compete with them; and, in the other great region, with those of the Old World. Consequently we have here everything favourable for much modification,--for far more modification than with the Alpine productions left isolated, within a much more recent period, on the several mountain ranges and on the arctic lands of Europe and N. America. Hence it has come, that when we compare the now living productions of the temperate regions of the New and Old Worlds, we find very few identical species; but we find in every class many forms, which some naturalists rank as geographical races, and others as distinct species; and a host of closely allied or representative forms which are ranked by all naturalists as specifically distinct[62]."

[60] Origin of Species, p. 332.

[61] Origin of Species, p. 332.

[62] Ibid. pp. 333-4.

In view then of all the above considerations--and especially those quoted from Darwin--it appears to me that far from raising any difficulty against the theory of evolution, the facts adduced by Mr. Carruthers make in favour of it. For when once these facts are taken in connection with the others above mentioned, they serve to complete the correspondence between degrees of modification with degrees of time on the one hand, and with degrees of evolution, of change of environment, &c., on the other. Or, in the words of Le Conte, when dealing with this very subject, "It is impossible to conceive a more beautiful illustration of the principles we have been trying to enforce[63]."

[63] Evolution and its Relation to Religious Thought, p. 194.

NOTE A TO PAGE 257.

The passages in Dr. Whewell's writings, to which allusion is here made, are somewhat too long to be quoted in the text. But as I think they deserved to be given, I will here reprint a letter which I wrote to Nature in March, 1888.

In his essay on the Reception of the Origin of Species, Prof. Huxley writes:--

"It is interesting to observe that the possibility of a fifth alternative, in addition to the four he has stated, has not dawned upon Dr. Whewell's mind" (Life and Lectures of Charles Darwin, vol. ii, p. 195).

And again, in the article Science, supplied to The Reign of Queen Victoria, he says:--

"Whewell had not the slightest suspicion of Darwin's main theorem, even as a logical possibility" (p 365).

Now, although it is true that no indication of such a logical possibility is to be met with in the History of the Inductive Sciences, there are several passages in the Bridgewater Treatise which show a glimmering idea of such a possibility. Of these the following are, perhaps, worth quoting. Speaking of the adaptation of the period of flowering to the length of a year, he says:--

"Now such an adjustment must surely be accepted as a proof of design, exercised in the formation of the world. Why should the solar year be so long and no longer? or, this being such a length, why should the vegetable cycle be exactly of the same length? Can this be chance?... And, if not by chance, how otherwise could such a coincidence occur than by an intentional adjustment of these two things to one another; by a selection of such an organization in plants as would fit them to the earth on which they were to grow; by an adaptation of construction to conditions; of the scale of construction to the scale of conditions? It cannot be accepted as an explanation of this fact in the economy of plants, that it is necessary to their existence; that no plants could possibly have subsisted, and come down to us, except those which were thus suited to their place on the earth. This is true; but it does not at all remove the necessity of recurring to design as the origin of the construction by which the existence and continuance of plants is made possible. A watch could not go unless there were the most exact adjustment in the forms and positions of its wheels; yet no one would accept it as an explanation of the origin of such forms and positions that the watch would not go if these were other than they were. If the objector were to suppose that plants were originally fitted to years of various lengths, and that such only have survived to the present time as had a cycle of a length equal to our present year, or one which could be accommodated to it, we should reply that the assumption is too gratuitous and extravagant to require much consideration."

Again, with regard to "the diurnal period," he adds:--

"Any supposition that the astronomical cycle has occasioned the physiological one, that the structure of plants has been brought to be what it is by the action of external causes, or that such plants as could not accommodate themselves to the existing day have perished, would be not only an arbitrary and baseless assumption, but, moreover, useless for the purposes of explanation which it professes, as we have noticed of a similar supposition with respect to the annual cycle."

Of course these passages in no way make against Mr. Huxley's allusions to Dr. Whewell's writings in proof that, until the publication of the Origin of Species, the "main theorem" of this work had not dawned on any other mind, save that of Mr. Wallace. But these passages show, even more emphatically than total silence with regard to the principle of survival could have done, the real distance which at that time separated the minds of thinking men from all that was wrapped up in this principle. For they show that Dr. Whewell, even after he had obtained a glimpse of the principle "as a logical possibility," only saw in it an "arbitrary and baseless assumption." Moreover, the passages show a remarkable juxtaposition of the very terms in which the theory of natural selection was afterwards formulated. Indeed, if we strike out the one word "intentional" (which conveys the preconceived idea of the writer, and thus prevented him from doing justice to any naturalistic view), all the following parts of the above quotations might be supposed to have been written by a Darwinian. "If not by chance, how otherwise could such a coincidence occur, than by an adjustment of these two things to one another; by a selection of such an organization in plants as would fit them to the earth on which they were to grow; by an adaptation of construction to conditions; of the scale of construction to the scale of conditions?" Yet he immediately goes on to say: "If the objector were to suppose that plants were originally fitted to years of various lengths, and that such only have survived to the present time ... as could be accommodated to it (i. e. the actual cycle), we should reply that the assumption is too gratuitous and extravagant to require much consideration." Was there ever a more curious exhibition of failure to perceive the importance of a "logical possibility"? And this at the very time when another mind was bestowing twenty years of labour on its "consideration."

NOTE B TO PAGE 295.

Since these remarks were delivered in my lectures as here printed, Mr. Mivart has alluded to the subject in the following and precisely opposite sense:--

Many of the more noteworthy instincts lead us from manifestations of purpose directed to the maintenance of the individual, to no less plain manifestations of a purpose directed to the preservation of the race. But a careful study of the interrelations and interdependencies which exist between the various orders of creatures inhabiting this planet shows us yet a more noteworthy teleology--the existence of whole orders of such creatures being directed to the service of other orders in various degrees of subordination and augmentation respectively. This study reveals to us, as a fact, the enchainment of all the various orders of creatures in a hierarchy of activities, in harmony with what we might expect to find in a world the outcome of a First Cause possessed of intelligence and will[64].

[64] On Truth, p. 493.

Having read this much, a Darwinian is naturally led to expect that Mr. Mivart is about to offer some examples of instincts or structures exemplifying what in the margin he calls the "Hierarchy of Ministrations." Yet the only facts he proceeds to adduce are the sufficiently obvious facts, that the inorganic world existed before the organic, plants before herbivorous animals, these before carnivorous, and so on: that is to say, everywhere the conditions to the occurrence of any given stage of evolution preceded such occurrence, as it is obvious that they must, if, as of course it is not denied, the possibility of such occurrence depended on the precedence of such conditions. Now, it is surely obvious that such a "hierarchy of ministrations" as this, far from telling against the theory of natural selection, is the very thing which tells most in its favour. The fact that animals, for instance, only appeared upon the earth after there were plants for them to feed upon, is clearly a necessity of the case, whether or not there was any design in the matter. Such "ministrations," therefore, as plant-organisms yield to animal-organisms is just the kind of ministration that the theory of natural selection requires. Thus far, then, both the theories--natural selection and super-natural design-- have an equal right to appropriate the facts. But now, if in no one instance can it be shown that the ministration of plant-life to animal-life is of such a kind as to subserve the interests of animal-life without at the same time subserving those of the plant-life itself, then the fact makes wholly in favour of the naturalistic explanation of such ministration as appears. If any plants had presented any characters pointing prospectively to needs of animals without primarily ministering to their own, then, indeed, there would have been no room for the theory of natural selection. But as this can nowhere be alleged, the theory of natural selection finds all the facts to be exactly as it requires them to be: such ministration as plants yield to animals becomes so much evidence of natural selection having slowly formed the animals to appropriate the nutrition which the plants had previously gathered--and gathered under the previous influence of natural selection acting on themselves entirely for their own sakes. Therefore I say it is painfully manifest that "the enchainment of all the various orders of creatures in a hierarchy of activities," is not "in harmony with what we might expect to find in a world the outcome of a First Cause possessed of intelligence and [beneficent] will." So far as any argument from such "enchainment" reaches, it makes entirely against the view which Mr. Mivart is advocating. In point of fact, there is a total absence of any such "ministration" by one "order of creatures" to the needs of any other order, as the beneficent design theory would necessarily expect; while such ministration as actually does obtain is exactly and universally the kind which the naturalistic theory requires.

Again, quite independently, and still more recently, Mr. Mivart alluded in Nature (vol. xli, p. 41) to the difficulty which the apparently exceptional case of gall-formation presents to the theory of natural selection. Therefore I supplied (vol. xli, p. 80) the suggestion given in the text, viz. that although it appears impossible that the sometimes remarkably elaborate and adaptive structures of galls can be due to natural selection acting directly on the plants themselves--seeing that the adaptation has reference to the needs of their parasites--it is quite possible that the phenomena may be due to natural selection acting indirectly on the plants, by always preserving those individual insects (and larvae) the character of whose secretions is such as will best induce the particular shapes of galls that are required. Several other correspondents took part in the discussion, and most of them accepted the above explanation. Mr. T. D. A. Cockerell, however, advanced another and very ingenious hypothesis, showing that there is certainly one conceivable way in which natural selection might have produced all the phenomena of gall-formation by acting directly on the plants themselves[65]. Subsequently Mr. Cockerell published another paper upon the subject, stating his views at greater length. The following is the substance of his theory as there presented:--

[65] Nature, vol. xli, p. 344.

Doubtless there were internal plant-feeding larvae before there were galls: and, indeed, we have geological evidence that boring insects date very far back indeed. The primitive internal feeders, then, were miners in the roots, stems, twigs, or leaves, such as occur very commonly at the present day. These miners are excessively harmful to plant-life, and form a class of the most destructive insect-pests known to the farmer: they frequently cause the death of the whole or part of the plant attacked. Now, we may suppose that the secretions of certain of these insects caused a swelling to appear where the larvae lived, and on this excrescence the larvae fed. It is easy to see that the greater the excrescence, and the greater the tendency of the larvae to feed upon it, instead of destroying the vital tissues, the smaller is the amount of harm to the plant. Now the continued life and vitality of the plant is beneficial to the larvae, and the larger or more perfect the gall, the greater the amount of available food. Hence natural selection will have preserved and accumulated the gall-forming tendencies, as not only beneficial to the larvae, but as a means whereby the larvae can feed with least harm to the plant. So far from being developed for the exclusive benefit of the larvae, it is easy to see that, allowing a tendency to gall-formation, natural selection would have developed galls exclusively for the benefit of the plants, so that they might suffer a minimum of harm from the unavoidable attacks of insects.

But here it may be questioned--have we proof that internal feeders tend to form galls? In answer to this I would point out that gall-formation is a peculiar feature, and cannot be expected to arise in every group of internal feeders. But I think we can afford sufficient proof that wherever it has arisen it has been preserved; and further, that even the highly complex forms of galls are evolved from forms so simple that we hesitate to call them galls at all[66].

[66] Entomologist, March, 1890.

The paper then proceeds to give a number of individual cases. No doubt the principal objection to which Mr. Cockerell's hypothesis is open is one that was pointed out by Herr Wetterhan, viz. "the much greater facility afforded to the indirect action through insects, by the enormously more rapid succession of generations with the latter than with many of their vegetable hosts--oaks above all[67]." This difficulty, however, Mr. Cockerell believes maybe surmounted by the consideration that a growing plant need not be regarded as a single individual, but rather as an assemblage of such[68].

[67] Nature, vol. xli, p. 394.

[68] Ibid. vol. xli, pp. 559-560.

NOTE C TO PAGE 394.

The only remarks that Mr. Wallace has to offer on the pattern of colours, as distinguished from a mere brilliancy of colour, are added as an afterthought suggested to him by the late Mr. Alfred Tylor's book on Colouration of Animals and Plants (1886). But, in the first place, it appears to me that Mr. Wallace has formed an altogether extravagant estimate of the value of this work. For the object of the work is to show, "that diversified colouration follows the chief lines of structure, and changes at points, such as the joints, where function changes." Now, in publishing this generalization, Mr. Tylor--who was not a naturalist--took only a very limited view of the facts. When applied to the animal kingdom as a whole, the theory is worthless; and even within the limits of mammals, birds, and insects--which are the classes to which Mr. Tylor mainly applies it--there are vastly more facts to negative than to support it. This may be at once made apparent by the following brief quotation from Prof. Lloyd Morgan:--

It can hardly be maintained that the theory affords us any adequate explanation of the specific colour-tints of the humming-birds, or the pheasants, or the Papilionidae among butterflies. If, as Mr. Wallace argues, the immense tufts of golden plumage in the bird of paradise owe their origin to the fact that they are attached just above the point where the arteries and nerves for the supply of the pectoral muscles leave the interior of the body--and the physiological rationale is not altogether obvious,--are there no other birds in which similar arteries and nerves are found in a similar position? Why have these no similar tufts? And why, in the birds of paradise themselves, does it require four years ere these nervous and arterial influences take effect upon the plumage? Finally, one would inquire how the colour is determined and held constant in each species. The difficulty of the Tylor-Wallace view, even as a matter of origin, is especially great in those numerous cases in which the colour is determined by delicate lines, thin plates, or thin films of air or fluid. Mr. Poulton, who takes a similar line of argument in his Colours of Animals (p. 326), lays special stress on the production of white (pp. 201-202).

As regards the latter point, it may be noticed that not in any part of his writings, so far as I can find, does Mr. Wallace allude to the highly important fact of colours in animals being so largely due to these purely physical causes.

Everywhere he argues as if colours were universally due to pigments; and in my opinion this unaccountable oversight is the gravest defect in Mr. Wallace's treatment both of the facts and the philosophy of colouration in the animal kingdom. For instance, as regards the particular case of sexual colouration, the oversight has prevented him from perceiving that his theory of "brilliancy" as due to "a surplus of vital energy," is not so much as logically possible in what must constitute at least one good half of the facts to which he applies it--unless he shows that there is some connection between vital energy and the development of striations, imprisonment of air-bubbles, &c. But any such connection--so essentially important for his theory--he does not even attempt to show. Lastly, and quite apart from these remarkable oversights, even if Mr. Tylor's hypothesis were as reasonable and well-sustained as it is fanciful and inadequate, still it could not apply to sexual colouration: it could apply only to colouration as affected by physiological functions common to both sexes. Yet it is in order to furnish a "preferable substitute" for Mr. Darwin's theory of sexual colouration, that Mr. Wallace adduces the hypothesis in question as one of "great weight"! In this matter, therefore, I entirely agree with Poulton and Lloyd Morgan.

###